Ibtissam Medarhri

Applications of Numerical Methods

Ibtissam Medarhri

Applications of Numerical Methods

Applications of Numerical Methods in Finance and Environmental Sciences

Noor Publishing

Imprint

Cover image: www.ingimage.com

Publisher:
Noor Publishing
is a trademark of
Dodo Books Indian Ocean Ltd. and OmniScriptum S.R.L publishing group

120 High Road, East Finchley, London, N2 9ED, United Kingdom
Str. Armeneasca 28/1, office 1, Chisinau MD-2012, Republic of Moldova, Europe
Managing Directors: Ieva Konstantinova, Victoria Ursu
info@omniscriptum.com

Printed at: see last page
ISBN: 978-620-9-08149-1

ABSTRACT:

This book explores two major areas of application of **numerical methods** in **finance** and **environmental sciences.**

In the first part, we address two financial problems: the *calibration of local volatility* from observed market option prices, and the *modeling of stochastic volatility.* Since the calibration process constitutes an *ill-posed inverse problem*, we propose an approach based on *Tikhonov regularization*, using *Dupire's partial differential equation* to model option prices. Furthermore, we develop *efficient numerical schemes* such as the *Direct Discontinuous Galerkin method*, the *explicit Euler method*, and *Runge–Kutta schemes of order two and three* for time discretization.

The second part focuses on *environmental applications*, addressing two types of problems. The first concerns the *identification of a source term* in a *saltwater intrusion problem*, which plays a key role in the sustainable management of water resources. The goal is to determine an unknown spatially dependent source from boundary data. The mathematical formulation relies on *variable separation* and *Tikhonov regularization* to overcome the ill-posedness of the problem. Numerical results were validated under *pumping* and *freshwater injection* scenarios.

The second problem analyzes a *new mixed formulation*, obtained by rewriting the system in matrix form and introducing an auxiliary variable

$$\sigma = R(u)\nabla u,$$

representing the *flux tensor* of the primary variable

$$u = (h, \Phi_f)^T.$$

Here, h denotes the *depth of the salt–freshwater interface*, Φ_f the *hydraulic head* of the freshwater, and $R(u)$ a *symmetric positive-definite diffusion matrix*. We show that both the continuous problem and its *mixed finite element approximation* are well-posed, and we provide *optimal error estimates* for both the steady and evolutionary cases.

Key words: Locale Volatility, Pricing options, Inverse problem, Regularization of Tikhonov, Discontinuous Galerkin method, Saltwater intrusion, Source identification, Mixed finite elements.

Notation and Symbols

- **General notation**

$\mathbb{R}^n$	n-dimensional Euclidean space.
$\mathbb{R}^+$	the non-negative real numbers.
$x = \{x_1, ..., x_n\}$	dx is defined as follows $dx = dx_1 ... dx_n$
A^T	the transposed of the matrix A.
x^+	$\max(0, x)$.
W_t	Brownian motion.
$\widehat{u}$	numerical flux.
$\lVert\cdot\rVert_\alpha$	is the norm in L_α.
$\mathcal{L}$	Infinitesimal generator.
CEV	Constant Elasticity of Variance.
DDG	Direct Discontinuous Galerkin.
DGM	Discontinuous Galerkin method.
FD	Finite Differences method.
FE	Finite Elements method.
FV	Finite Volume method.
PDE	Partial Differential Equation.
PIDE	Partial Integro-Differential Equation.
RKDG	Runge-Kutta Discontinuous Galerkin.
RMS	Root Mean Squared.
SDE	Stochastic Differential Equation.

Contents

List of Figures

List of Tables

List of Algorithms

CHAPTER 1
Introduction

≪ *All models are wrong, but some are useful.*≫
- George Box

Applied mathematics has advanced through the joint action of (i) an internal drive to create and diversify mathematical tools, and (ii) continuous enrichment from neighboring sciences and technologies (physics, chemistry, computer science, biology, and industry), as well as economics, finance, and the social sciences. This book illustrates two complementary fields of application: **quantitative finance** and **hydro-environmental modeling**, connected by a common thread: modeling, *inverse problems*, and *numerical methods*.

Financial mathematics in brief: history, models, and calibration

Financial mathematics, an integral part of applied mathematics, has an original status and a particular history. The growing interest stems from both deep mathematical questions (stochastic processes, probabilities, PDEs with stochastic coefficients, statistics) and the sustained effort of banks and insurers to fund research on pricing and risk for *derivatives* [79, 83, 71, 86].

Historically, derivatives predate modern exchanges. Thales (VIIth century BC) used options to secure the right to olive presses ahead of harvest, illustrating the logic of *rights without obligations*. Forward contracts hedged farmers and enabled speculation long before standardized markets. The modern era began with the CBOE (1973), introducing standardized option contracts. A derivative is a contract whose value depends on an underlying (equity, bond, commodity, index) and comes in families such as forwards/futures, swaps, and options [79].

The quantitative lineage goes from Bachelier?s [15] to the calculus of Kolmogorov [90] and Itô [80], culminating in the Black–Scholes–Merton framework [22, 100]. Pricing/hedging is formulated via SDE/PDE, with Feynman–Kac links and no-arbitrage arguments.

Calibration as an inverse problem. Practically, models contain latent quantities (e.g., local or stochastic volatility) that must be *calibrated* from noisy, incomplete market data. A classical formulation minimizes

$$\min_{\sigma} F(\sigma) := \|V(\sigma) - \widetilde{V}\|^2, \tag{1.1}$$

where $\widetilde{V}$ are observed option prices and $V(\sigma)$ are model-implied prices computed via a PDE (e.g., Dupire). Because the problem is underdetermined and ill-posed, Tikhonov-type regularization is essential [93, 34, 25, 50, 81].

Numerical solvers. We use high-order PDE engines for pricing and Greeks: discontinuous Galerkin in space coupled with Runge–Kutta in time (RKDG), alongside finite-difference schemes and stochastic Runge–Kutta for stochastic volatility models (e.g., CEV).

Environmental modeling: saltwater intrusion in coastal aquifers

Freshwater is unevenly distributed and critically stressed in many regions despite continual renewal by the hydrologic cycle [18]. Baseline water stress is projected to intensify in coming decades [98]. In Morocco, recent assessments indicate *acute water stress*, particularly in densely populated, agriculture-oriented regions such as Marrakech–Safi, Casablanca–Rabat and Ksar El Kebir [77]. Overuse of strategic aquifers and semi-arid climate regimes aggravate scarcity.

Saltwater intrusion arises from the migration of saline fronts into coastal aquifers, degrading resources via freshwater-level decline and salinization. Case studies report marine intrusion near El Jadida under overexploitation [126], high vulnerability at Mnasra amplified by pumping [16], and advancing saline wedges with upconing at Foum El Oued (Laayoune) [64].

Modeling frameworks. Since Ghyben–Herzberg [122, 107] and Reilly–Goodman [110, 111], key mechanisms include variable density, dispersion and the flow regime above the saline wedge [17, 20, 19, 18]. Two main lines coexist:

- *Sharp interface (non-miscible)*: a moving, abrupt front separates fresh and saline waters (no mass transfer). Mathematically simpler; theory and numerics in [63, 104, 38, 99, 82, 13]. In unconfined settings, maximum-principle and interface-ordering issues may arise [99, 37].
- *Miscible flow*: co-flow with a finite mixing zone due to dispersion; more realistic but computationally demanding [12, 89, 131, 72, 35, 5].

To reconcile simplicity and realism, we adopt the *mixed sharp/diffuse-interface* approach [36], inspired by phase-field ideas (Allen–Cahn). It removes degeneracies, restores realistic maximum principles in unconfined aquifers [3], provides gradient-regularity and uniqueness [39], and has effective numerics [8, 33].

Inverse problems and data assimilation. As in finance, groundwater management hinges on inverse problems under noise: estimating hydraulic parameters or *source terms* (location/intensity). For sharp interfaces in confined aquifers, see [1, 118]; for the mixed approach in unconfined settings, [4]. Of special interest is source identification with separable structures $F(x)H(t)$ and boundary measurements [91]; related methods exist for heat/advection–diffusion [66, 67, 91] and wave models [103]. Stability (Hadamard) issues [87] motivate regularization and prior constraints.

Mixed finite elements (MFEM). Accurate fluxes and *local mass conservation* are crucial. MFEM approximate primary variables and fluxes simultaneously, with strong conservation properties [23, 29, 40, 69, 108]. In saturated media, theory and comparisons with FV/FD show advantages for heterogeneous formations [31, 61, 102, 85, 125]. In unsaturated settings, DG/SPDMEM combinations achieve optimal L^2 error estimates [96, 127, 128].

Methodological outlook of this book

The book is organized around three methodological pillars that interplay across finance and environment:

1. **Inverse modeling and calibration.** We study local-volatility calibration via (1.1), using Tikhonov regularization and efficient PDE solvers. Following [93, 34, 25], we emphasize robust discretizations and practical regularizers that reconcile data fit and stability. In parallel, we formulate and solve inverse source problems in coastal aquifers (e.g., reconstructing

$F(x)$ in a separable source $F(x)H(t)$), leveraging linearization, decomposition of boundary conditions, and stability estimates.

2. **Forward solvers and numerical schemes.** For option pricing, we employ discontinuous Galerkin in space and Runge–Kutta in time (RKDG) for European/American contracts, and we also treat stochastic volatility models (e.g., CEV). For aquifers, we adopt the mixed sharp/diffuse-interface approach [36], with MFEM formulations that provide local mass conservation and optimal L^2 error estimates for both states and fluxes.

3. **Analysis, well-posedness and error control.** We present streamlined analyses: (i) existence/uniqueness and maximum principles where applicable; (ii) stability of inverse reconstructions under noise; (iii) a priori error estimates for the chosen schemes (DG/MFEM/FD), with attention to degeneracies, nonlinearity and coupling in the environmental models.

Chapter 2
Preliminary

≪ *Suppose we use the standard deviation of possible future returns on a stock as a measure of its volatility. Is it reasonable to take that volatility as a constant over time? I think not.*≫

- Fisher Black

Contents

2.1 Finance

The development of mathematical methods for pricing derivatives was a major reason for the expansion of derivative markets. Such theoretical achievement was recognized by the Nobel Prize in Economics award to **R. Merton** and **M. Scholes**, the corresponding methods involve the solution of the Black-Scholes partial differential equation.(see [56])

The right price of an option is determined by a number of factors, including the volatility of the underlying asset. Unfortunately, this parameter can not directly read off of the market data. But the essential question is: What is volatility and why is it so important?

The concept of the volatility calibration is probably the central subject that attract the most researchers in the area of mathematical finance. This interest is motivated by two main reasons: more and more companies using the tools of risk management, and the large number of derivatives traded in global financial markets.

First of all, we present the generalities about options, and the different types of volatility. Then, we will focus on recent work for calibrating the volatility function.

2.1.1 Option Pricing

2.1.1.1 The Stock Price Process

The relationship between the stock market and the mathematical concept of Brownian Motion goes back to Bachelier. A Brownian Motion corresponds to a process the increments of which are independent stationary normal random variables. Given that a Brownian Motion can take negative values, it cannot be used for stock price.

In other words, the stock price S evolves according to a Stochastic Differential Equation (**SDE**) of the form

$$dS_t = \mu S_t dt + \sigma S_t dW_t \tag{2.1}$$

where dW_t is a Brownian Motion process, μ is the drift (the instantaneous expected total return of the stock) and σ the instantaneous standard deviation of stock price returns, called the *volatility* in financial markets.

In the following we will recall Itô's lemma, given in [79], which is required in order to write the deterministic model using a stochastic one.

Lemma 1.

Itô's lemma, states the following:

$$dx(t) = \mu_t dt + \sigma_t dB_t$$

and any twice differentiable scalar function f(t,x) of two real variables t and x, one has

$$df(t,x) = \frac{\partial f}{\partial t}dt + \frac{\partial f}{\partial x}dX_t + \frac{\sigma_t^2}{2}\frac{\partial^2 f}{\partial x^2}dt.$$

Using Itô's lemma, we obtain

$$\begin{aligned} d(\ln S_t) &= \frac{1}{S_t} dS_t + \frac{1}{2}(-\frac{1}{S_t^2})(\sigma S_t)^2 dt, \\ &= \frac{1}{S_t}(\mu S_t dt + \sigma S_t dW_t) - \frac{1}{2}\sigma^2 dt \end{aligned}$$

Then,

$$d\ln(S_t) = (\mu - \frac{1}{2}\sigma^2)dt + \sigma dW_t \tag{2.2}$$

The Stock return μ could easily becomes time dependent without changing any of our arguments.
expressed formally, a variable W_t follows a Wiener process it it has the following property.

Definition 1. *The change ΔW_t during a small period of time Δt is*

$$\Delta W_t = \varepsilon\sqrt{\Delta t}$$

where ε has a standard normal distribution $\phi(0,1)$.

The equation (2.1) represents a continuous process. We can either take this as an approximation to the real discrete tick by tick stock movements, or consider it the real unobservable dynamics of the stocks price, in which case the discrete prices constitute a sample from this continuous ideal process.
Either way, the use of a continuous equation makes the pricing of financial instruments more analytically tractable.

Using the definition 1, the discrete form of (2.2) (for more details see[79]) is:

$$\ln(S_{t+\Delta t}) = \ln(S_t) + (\mu - \frac{1}{2}\sigma^2)\Delta t + \sigma\varepsilon\sqrt{\Delta t} \tag{2.3}$$

where ε has a standard normal distribution.

2.1.2 Definition and generalities of options contracts

A derivative is an instrument which derives its value from another entity known as "underlying" such as share, index, interest rate, currency,... etc. Derivatives are one of the three main categories of financial instruments, the other two being equities (i.e. stocks) and debt (i.e. bonds and mortgages). Derivatives include a variety of financial contracts, including futures, forwards, swaps, options, and variations of these such as caps, floors, collars, and credit default swaps.

Derivatives are a contract between two parties that specify conditions (especially the dates, resulting values and definitions of the underlying variables, the parties,contractual obligations, and the notional amount) under which payments are to be made between the parties. The most common underlying assets include commodities, stocks, bonds, interest rates and currencies, but they can also be other derivatives, which add another layer of complexity to proper valuation. (given in ***"Investment Banking Products"*** [92])

An option is a financial contract dependent on an underlying. It provides the holder with the right, and not an obligation, to buy or sell a specified quantity of an underlying asset at a fixed price (called a Strike) at or before the expiration date of the option.
There are two basic types of options: calls and puts. A **Call** option gives the holder the right to buy a specific number of the underlying asset at a certain price on a certain date or during a period. A **Put** option gives the holder the right to sell a specific number of the underlying asset at a certain price on a certain date or during a period.

The type of option determines the period during which the right of the buyer of the option may be exercised: (the definitions are given in [79])

- **European Option**: the right can be exercised only on the expiration date.
- **American Option**: the right can be exercised at any time up to the expiration date.

There are many other options available on the market, called exotic options among its options there are: (the definitions are given in [79, 86, 83])

- **Asiatic Option**: the right can be exercised only on the expiration date. The payoff is dependent of the average price of the underlying asset, over some prespecified period and prespecified frequency, during the lifetime of the option. The arithmetic asian call option uses the ordinary arithmetic mean, $\widehat{x}(t) = \frac{1}{t-a}\int_a^t x(s)ds$, and has the following payout function, $V(t) = (\frac{1}{t-a}\int_a^t S(s)ds - K)^+$ and a denotes the point in time from where the mean is taken.

- **Digital Option:** or binary option pays out a unit amount on the maturity date if the spot of the underlying asset is above some predetermined boundary. The payoff function is $V(t) = 1_{S_t \geq B}$, where $1_{\{A\}}$ is the indicator function, which takes value 1 if the event A occurs, and 0 otherwise.

- **Barrier Option**: the payoff depends in whether the underlying asst's price reaches a certain level during a certain period of time (down-and-in/out, up-and-in/out,...). The spot price trajectory stays above the boundary during the whole time interval the option pays out 1. Formally we have, $V(t) = 1_{\left\{\inf_{t^* \in [0,t]} S^*_t > B\right\}}$. Similar to down-and-out is the up-and-out digital option.

There are two sides to every option contract: the long position and short position. The holder having bought the option has taken the long position. The holder having sold the option has taken the short position. The buyer of the option pays a premium to gain the right to buy the underlying asset. The seller of the option receives the premium and then bears the obligation to sell the underlying asset. The writer's profit or loss is opposite to that for the purchaser of the option.

Assume S is the final price of the underlying asset, X is the exercise price, and P is the premium. The payoff of a long position on a European call at the expiration date is $\max(S-X,0)$ for the option will be exercised only when $S > X$. Next figures illustrate payoff diagrams [79].

The profit for a long position on a European call in 2.1 (a) is

$$\max(S-X,0)-P$$

The profit for a short position on a European call in 2.1 (b) is

$$-(\max(S-X,0)-P)=\min(X-S,0)+P$$

The profit for a long position on a European put in 2.1 (c) is

$$\max(X-S,0)-P$$

The profit for a short position on a European put in 2.1 (d) is

$$-(\max(X-S,0)-P)=\min(S-X,0)+P$$

2.1.3 Black-Scholes-Merton Model

The F. Black & M. Scholes (1973) made a major revolutionary in pricing option by developing the well-know Black-Scholes model for calculating the premium of an European call and put options.

The assumptions used to derive the Black-Scholes differential equation are:

1. The stock price follows the geometric Brownian motion with μ and σ constant,
2. The short selling of securities with full use of proceeds is permitted,
3. There are no transactions costs or taxes; all securities are perfectly divisible,
4. There are no dividends during the life of the derivative,

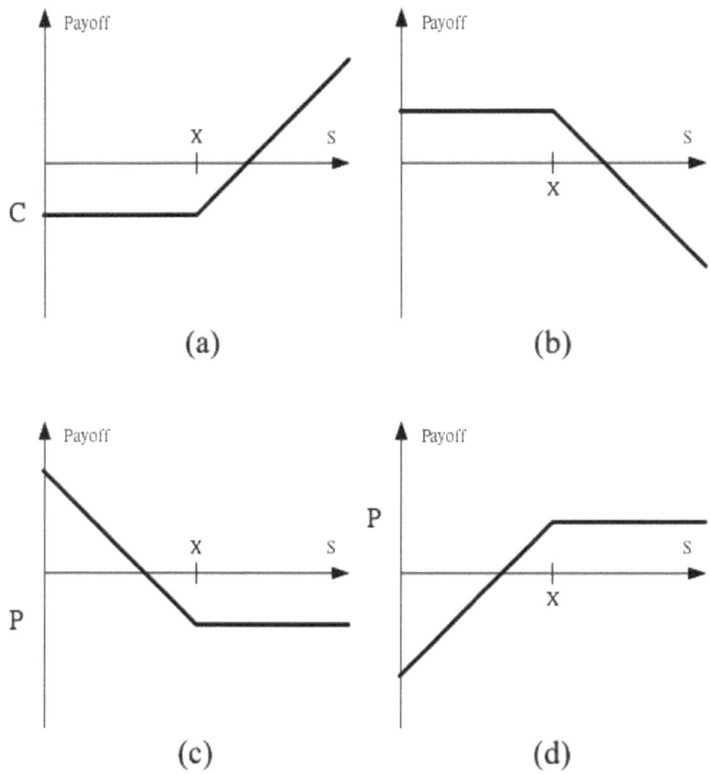

Figure 2.1: Option payoff: a) long call b) short call c) long put d) short put

5. There are no riskless arbitrage opportunities,

6. Security trading is continuous,

7. The risk-free rate of interest, r, is constant and the same for all maturities.

2.1.3.1 The Black-Scholes Differential Equation

Under the assumptions listed above, Black & Scholes form a riskless portfolio consisting of a position in the option and a position in the underlying stock.
Suppose that V is the price of a call option or other derivative contingent on S. The

variable V must be some function of S and t. The transition from a stochastic equation EDP is a deterministic EDP performed using Ito's lemma.

$$\Delta V = (\frac{\partial V}{\partial S}\mu S + \frac{\partial V}{\partial t} + \frac{1}{2}\frac{\partial^2 V}{\partial S^2}\sigma^2 S^2)\Delta t + \frac{\partial V}{\partial S}\sigma S \Delta w \tag{2.4}$$

where ΔV and ΔS are the changes in V and S in a small time interval Δt, and w is the Wiener process.
The holder of the portfolio is short one derivative and long an amount $\partial V/\partial S$ of shares. Define Π as the value of the portfolio. By definition

$$\Pi = -V + \frac{\partial V}{\partial S}S \tag{2.5}$$

The change $\Delta\Pi$ in the value of the portfolio in the time interval Δt is given by:

$$\Delta\Pi = -\Delta V + \frac{\partial V}{\partial S}\Delta S \tag{2.6}$$

Then, substituting equations (2.4), into equation (2.6) yields:

$$\Delta\Pi = \Big(- \frac{\partial V}{\partial t} - \frac{1}{2}\frac{\partial^2 V}{\partial S^2}\sigma^2 S^2 \Big)\Delta t \tag{2.7}$$

The assumptions listed for Black-Scholes model imply that the portfolio must instantaneously earn the same rate of return as other short-term risk-free securities. If it earned more than this return, arbitrageurs could make a riskless profit by borrowing money to buy the portfolio; if it earned less, they could make a riskless profit by shorting the portfolio and buying risk-free securities. It follows that

$$\Delta\Pi = r\Pi\Delta t \tag{2.8}$$

where r is the risk-free interest rate. Therefore, return from that portfolio should equal to risk-free interest rate (r), in the absence of arbitrage argument.
By replacing the terms Π and $\Delta\Pi$ of the equation (2.8) bye their values given in (2.5) and (2.8), is obtained

$$\frac{\partial V}{\partial t} + rS\frac{\partial V}{\partial S} + \frac{1}{2}\sigma^2 S^2 \frac{\partial^2 V}{\partial S^2} = rV \tag{2.9}$$

where V is the price of a derivative security, S is the stock price, σ is the volatility of the stock price. The equation (2.9) is the Black-Scholes differential equation. The solution of this PDE depends on the boundary conditions and the initial or terminal conditions which characterize the derived product considered.
In the case of a European call option, the terminal condition is

$$V = max(S - K; 0), \quad quand \quad t = T$$

In the case of a European put Option, when $t = T$, the terminal condition is written as follow:

$$V = \max(K - S, 0).$$

where S the underlying, and K the strike.

2.1.3.2 The Black-Scholes pricing formula

The *PDE* BS can then be solved for different pay-off function, is called the closed form formula for European options, it's developed by three economists F. Black, M. Scholes, and R. Merton. The Black-Scholes formulas for the prices at time zero of European call option and European put option is given as follow:

$$C = S\mathcal{N}(d_1) - Ke^{-rT}\mathcal{N}(d_2) \tag{2.10}$$

and

$$P = Ke^{-rT}\mathcal{N}(-d_2) - S\mathcal{N}(-d_1) \tag{2.11}$$

where

$$d_1 = \frac{1}{\sigma\sqrt{T}}\left[\ln(\frac{S}{K}) + (r + \frac{\sigma^2}{2})T\right]$$

$$d_2 = \frac{1}{\sigma\sqrt{T}}\left[\ln(\frac{S}{K}) + (r - \frac{\sigma^2}{2})T\right] = d_1 - \sigma\sqrt{T}$$

C is the price of a European call, P is the price of a European put. And $N(x)$ is the probability distribution function of standard normal distribution.

2.1.3.3 Limitation of the Black-Scholes model

Despite the importance of the Black-Scholes model in the world of finance, some assumptions are not validated in practice. Among them:

- The model implies that the log-returns are normally distributed. Empirical studies of actual market date have suggested that this is often not the case.
- It is a continuous model. Trading takes place in discrete time intervals.
- The model assumes zero transaction costs.
- The volatility is assumed to be known and constant over time. This assumption has been relaxed by a number of authors. We will present some models in this book.

2.1.3.4 Numerical methods for option pricing

2.1.3.5 Monte Carlo approach

The Monte Carlo approach is a method often used in mathematical finance to calculate the value of an option, usually used for derivatives where the payoff is dependent on the history of the underlying variable or where there are several underlying variables. (inspired from [97, 10, 79])

When used to value an option , Monte Carlo simulation uses the risk-neutral valuation result. Consider a derivative dependent on a single market variable S that provides a payoff at time T. Suppose that the process followed by the underlying market variable in a risk-neutral world is:

$$dS = \hat{\mu} S dt + \sigma S dz$$

where dz is a Wiener process, $\hat{\mu}$ is the expected return in a risk-neutral world, and σ is the volatility.
In practice, it is usually more accurate to simulate $\ln(S)$ rather than S. Moreover, for lognormal random walk there is simple, and exact, time stepping algorithm. Ito's lemma leads to:

$$d(\ln(S) = \left(\hat{\mu} - \frac{1}{2}\sigma^2\right) dt + \sigma \varphi \sqrt{dt}$$

which after integrating provides:

$$S(t + \Delta t) = S(t) \exp\left[(\hat{\mu} S(t) - \frac{\sigma^2}{2})\Delta t + \sigma \varepsilon \sqrt{\Delta t}\right] \tag{2.12}$$

where $S(t)$ denotes the value of S at time t, ε is a random sample from a normal distribution with mean zero and standard deviation of 1.0.

2.1.3.6 Binomial Tree

The binomial Tree method is the first numerical procedure for option pricing, it is one of the simplest and most widely used methods. The famous binomial tree method is of Cox, Ross Runbeistein (CRR) [53].
Binomial tree valuation approach involves dividing the life of the option into a large number of small time intervals of length Δt. It assumes that in each time interval the price of the underlying asset moves from its initial value of S to one of two new values, Su and Sd. The approach is illustrated in the figure as follows:

Risk-neutrality demands: $\exp(rdt) = pu + (1 - p)d$, p is probability of an up movement, and since it is binomial tree $(1 - p)$ is the probability of down movement.
Using condition introduced by (CRR):

$$u = \frac{1}{d}$$

Then we obtain,

$$p = \frac{a - d}{u - d}$$

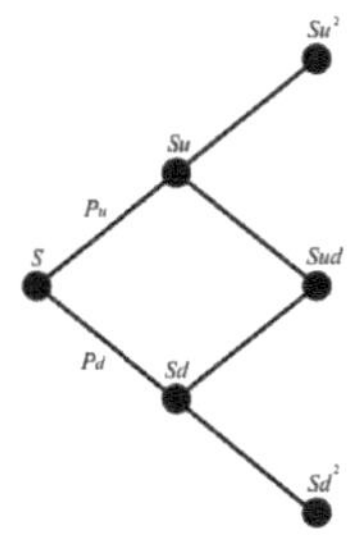

Figure 2.2: Binomial tree

$$u = \exp(\sigma\sqrt{\Delta t})$$
$$d = \exp(-\sigma\sqrt{\Delta t})$$

where

$$a = \exp(r - q)\Delta t$$

The variable a is sometimes referred to as the growth factor. Figure 2.2 illustrates the complete tree of asset prices that is considered when the binomial model is used. At time zero, the asset price, S_0, is known. At time Δt, there are two possible asset prices, S_0u and S_0d; at time $2\Delta t$, there are three possible asset prices, S_0u^2, S_0, and S_0d^2. In general, at time $i\Delta t$, we consider $i + 1$ asset prices. All feasible prices are obtained using:

$$S_0u^jd^{i-j}, \quad j = 0, 1, \cdots, i$$

The final step is at current time and stock price, where the theoretical fair value of the option is calculated. Algebraically it means for call option:

$$f_{N,j} = \max(S_0u^jd^{N-j} - K, 0), \quad j = 0, 1, ..., N$$

And for put option

$$f_{N,j} = \max(K - S_0u^jd^{N-j}, 0), \quad j = 0, 1, ..., N.$$

The advantage of binomial trees is easiness of implementation and it's easy to understand it.

2.1.3.7 Finite Difference Method

Finite difference (FD) method is the generic term for a large number of procedures that can be used for solving a (partial) differential equation that the derivative satisfies.

The FD method have as a common dominator some discretization scheme approaching required derivatives, and very useful for a small number dimension.

Idea in finite difference method is to find solution for differential equation by approximating every partial derivative numerically. FD method [101] create a mathematical relationship which links together every point on the solution domain. The boundary conditions are the first links in the chain, we discover what every point in the domain has to be.

The most popular FD methods used in computational finance are: *Explicit Euler, Implicit Euler, and Crank-Nicolson method.* Each of these three methods has its advantages and disadvantages. The Explicit Euler method is the easiest scheme to implement, but his main disadvantage is that it is unstable for certain choices of domain discretization. Implicit Euler and Crank-Nicolson are implicit methods, which generally require a resolution of a system of linear equations to be solved at each time step, which can be computationally intensive on a fine mesh, but they are unconditionally stable.

To illustrate the approach, we consider how it might be used to value the European put option [10, 79, 28, 112]. The differential equation that the option must satisfies the equation (2.9). We first define the domain discretization in S & t:

Suppose that S_{max} is a maximal price of underlying. We define $\Delta S = S_{max}/M$, and consider a total of $M+1$ equally spaced stock prices: $\{0, \Delta S, 2\Delta S, ..., S_{max}\}$.

Time discretization: $\{0, \Delta t, 2\Delta t, ..., N\Delta t\}$ where $\Delta t = T/N$, and $V_{i,j} = V(i\Delta S, j\Delta t)$, $i = 0, 1, 2, ..., M$, $j = 0, 1, 2, ..., N$.

The boundary conditions for the European Put are:

$$V(S,T) = \max(K - S, 0) \qquad V(0,t) = Ke^{-r(T-t)} \qquad V(S_{max}, t) = 0$$

The numerical form is written as follows:

$$V_{i,N} = \max(K - (i\delta S), 0) \qquad i = 0, 1, ..., M$$

$$V_{0,j} = Ke^{-r(N-j)\delta t} \qquad j = 0, 1, ..., N$$

$$V_{M,j} = 0 \qquad j = 0, 1, ..., N$$

1. Space Discretization: there are three Probabilities to approximate $\frac{\partial V}{\partial S}$
   (a) Backward difference : $\frac{\partial V}{\partial S} = \frac{V_i - V_{i-1}}{\Delta S}$
   (b) Central difference : $\frac{\partial V}{\partial S} = \frac{V_{i+1} - V_{i-1}}{2\Delta S}$

(c) Forward difference : $\frac{\partial V}{\partial S} = \frac{V_{i+1}-V_i}{\Delta S}$

The numerical approximation for $\frac{\partial^2 V}{\partial S^2}$ is generally given by:

$$\frac{\partial^2 V}{\partial S^2} = \frac{V_{i-1} - 2V_i + V_{i+1}}{\Delta S^2}, \qquad for \quad i = 1, 2, ..., M-1$$

2. Temporal discretization: Using the central difference with respect to stock price S, and using the explicit scheme in time. Since we are letting time flow backwards in the valuation procedure, the difference approximation becomes.

$$\frac{V_{i,j} - V_{i,j-1}}{\Delta t} + rS_i \frac{V_{i+1,j} - V_{i-1,j}}{2\Delta S} + \frac{1}{2}\sigma^2 S_i^2 \frac{V_{i+1,j} - 2V_{i,j} + V_{i-1,j}}{\Delta S^2} = rV_{i,j}$$

This equation can be write:

$$V_{i,j-1} = a_i^* V_{i-1,j} + b_i^* V_{i,j} + c_i^* V_{i+1,j}, \qquad j = N, N-1, ..., 1, 0; \quad i = 1, 2, ..., M-1$$

The prices at $j = N$ are known, $M - 1$ options values in $N - 1$-th step can be calculated easily using derived expression, other two values (at $i = 0$ and $i = N$) are known from boundaries.

2.1.3.8 Other Numerical Methods for Pricing Options

There are other numerical methods for Pricing Option, such as Finite Element Method (FEM), Finite Volume Method (FVM), and Discontinuous Galerkin Method (DGM),...

- Finite Element Method (FEM) [10, 79, 30, 112, 41]: was introduced by R. Courant [52] in 1943, for solving the partial differential equations arising in solid mechanics; the main idea was to find approximate solutions to boundary value problems for differential equations. It allows the passage of an infinite-dimensional problem defined on a Hilbert space V to a finite dimensional subspace V_h of V.
  The most popular reference on **FEM** in solid mechanics is the book of Zienkiewicz [130]. Generalizations to other fields of physics or engineering have been done by applied mathematicians through the concept of variational formulations and weak forms of the partial differential equations.

  In the simpler finite element methods, the construction of the approximation space V_h is done as follow:

  - The domain is partitioned into nonoverlapping cells (elements) whose shapes are simple and fixed: for example, intervals in one dimension, triangles or quadrilaterals in two dimensions, tetrahedra, or hexahedra in three dimensions.

  - The maximal degree k of the polynomial approximation in the elements is chosen.

  - V_h is made of functions of V whose restriction to the elements are polynomial of degree less than k.

- Finite Volume Method (FVM) [10, 79, 30, 112]: is a method for representing and evaluating partial differential equations in the form of algebraic equations (LeVeque [95] in 2002, and Toro [121] in 1999). It's a discretization method which is suitable for the numerical simulation of various types (elliptic, parabolic or hyperbolic, for instance) of conservation laws. Some of the important features of the finite volume method are similar to chose of the finite element method, it may be used on arbitrary geometries, using structured or unstructured meshes, and it leads to robust schemes.

  An additional feature is the local conservatively of the numerical fluxes, that is the numerical flux is conserved from one discretization cell to its neighbor. This last feature makes the finite volume method quite attractive when modeling problems for which the flux is of importance.
  The finite Volume method is locally conservative because it is based on a "balance" approach: a local balance is written on each discretization cell which is often called "control volume"; by the divergence formula an integral formulation of the fluxes over the boundary of the control volume is then obtained. In finance, they may be useful, for example for Asian options when the PDE becomes close to hyperbolic near maturity.

- Discontinuous Galerkin Method (DGM) [45, 48, 58, 75]: is somewhere between a finite element and a finite volume method and has many good features of both. An important distinction between the DG method and the usual finite-element method is that in the DG method the resulting equations are local to the generating element. The solution within each element is not reconstructed by looking to neighboring elements. This method is presented in chapter 4.

In this book we became interested on two numerical methods for option pricing. The first method used is the Finite Difference (FD) Method : in the seconde chapter for estimation Volatility, for the resolution of the inverse and the direct problem (pricing european option). In the fourth chapter the method is used for pricing European Option with the Constant Elasticity Variance (CEV) diffusion model. The second method is Discontinuous Galerkin Method (DGM): is using to compute the european and american option.

2.1.4 The Volatility Problem

The volatility is a statistical measure of the dispersion of returns for a given security or market index. The Volatility is a parameter, in option pricing formulas, showing the extent to which the return of the underlying asset will fluctuate between now and the option's expiration. Volatility, as expressed is a percentage coefficient within option pricing formulas, arises from daily trading activities. This section is based on the book of "**Volatility Articles**" [2].

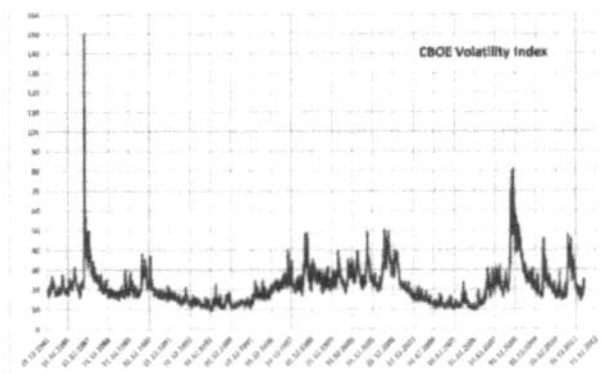

We present in the next figure different types of Volatility

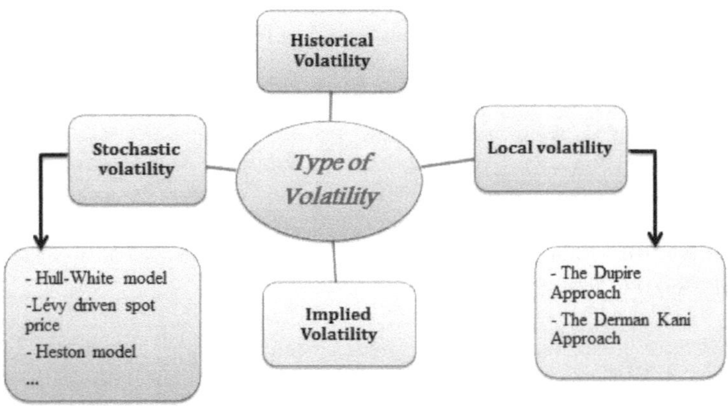

Figure 2.3: Different types of Volatility

2.1.4.1 Historical and implied Volatility

2.1.4.2 Historical Volatility

The term volatility has its roots within the Statistical field, but to get a more practical angle, the movements of stock prices are often used as a reference point for the definition of the volatility in a financial context (named *historical volatility*).

To estimate the volatility of a stock price empirically, the stock price is usually observed at fixed intervals of time (e.g, every day, week, or month, ...).

Considering N the number of observations, $S_0, S_1, ...S_N$ a sequence of known historic daily stock close prices, and τ the length of time interval in years, calling $R_n = \ln(S_{n+1}/S_n)$ the stock price return between two days and $\overline{R} = \frac{1}{N}\sum_{n=0}^{N-1} R_n$ the mean return, the historical volatility would be the annualized standard deviation of the returns, namely

$$\sigma_{hist} = \frac{\sqrt{\frac{1}{N-1}\sum_{n=0}^{N-1}(R_n - \overline{R})^2}}{\sqrt{\tau}} \tag{2.13}$$

Choosing an appropriate value for n is not easy. More data generally lead to more accuracy, but σ does change over time and data that are too old may not relevant for predicting the future volatility. A compromise that seems to work reasonably well is to use closing prices from daily data over the most recent 90 to 180 days. An alternative was suggested by Parkinson [106], where instead of daily close prices, we can use the high and the low prices of the stock on that day, and $R_n = \ln(S_n^{high}/S_n^{low})$.

The volatility would then be

$$\sigma_{parkinson} = \frac{\sqrt{\frac{1}{N-1}\frac{1}{4\ln(2)}\sum_{n=0}^{N-1}(R_n - \overline{R})^2}}{\sqrt{\tau}} \tag{2.14}$$

This second moment estimation derived by Parkinson, is based upon the fact that the range R_n of the asset, follows a feller distribution. One natural extension of the constant volatility approach is to make σ_t a deterministic function of time. This is equivalent to giving the volatility a term structure.

Much research effort has been devoted to the subject over the past years and several approaches and models have been proposed and evaluated across different markets and asset classes. Historical volatility is often used as the best predictor for future levels which makes it very important concepts to master. Popular models for forecasting volatility are among others ARCH and GARCH, and many more.

2.1.4.3 Implied Volatility

Even though future volatility is not directly observable in the market it's actually possible to extract the markets expectation of future volatility options traded on public exchanges. This estimate has been named implied volatility and is based on the Black-Scholes formula.

The Black-Scholes formula was used perfectly before the huge crashes of 1987 and 1989, but after, it has been observed that market prices contradict the assumption of constant volatility. The option prices for different maturities or strikes change drastically.

The question is: *What volatility is implied in observed option prices, if the Black-Scholes model is a valid description of market conditions?* Thus, the **implied volatility** is an indicator of the current sentiment of the market. This is the value of σ such that the price of an European Call (or put) is equal to the value obtained by applying the Black-Scholes formula.

The implied volatility is different from **historical volatility** because the latter is calculated from known past returns of an option. On the market, the implied volatility is that volatility which use the traders and is used to list underlying.

Historical Volatility	Implied Volatility
←	→
Looks back in time to show where volatility has been in the past.	Trader's view of expected future volatility based on current option prices.
Shows expected trading range of market	Indicator of the current sentiment of the market.

Table 2.1: Different between Implied and Historical Volatility

According to the model, we should expect implied volatilities σ_{imp} for different calls to be identical. But it can be observed that this is not the case: there is a dependence on the strike price K, on the moneyness and on the remaining life of the option. The Black-Scholes model doesn't describe the market correctly; one denote this effect the **Smile**[1]. The surface of volatility smile, changes shape from day to day, but some general features persist. In the next figure, we see the graph which explain the notation smile.

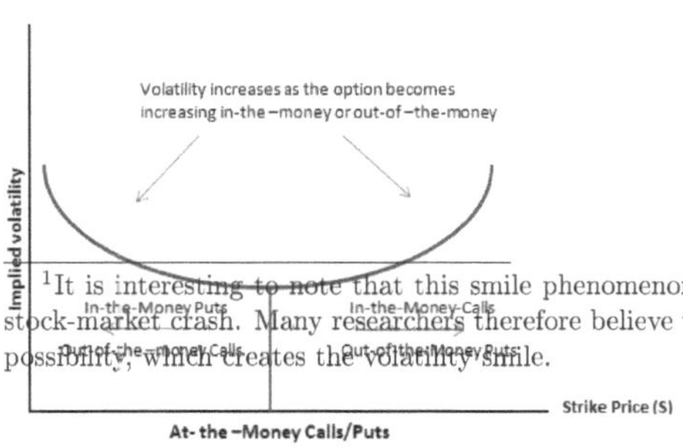

Figure 2.4: Smile-effect

When the implied volatility is relatively low, the option is considered cheap. When the implied volatility is relatively low and is excepted to rise, the best time to buy options. When the implied volatility is rela-

[1]It is interesting to note that this smile phenomenon was practically non-existent prior to the 1987 stock-market crash. Many researchers therefore believe that the markets have learnt to factor-in a crash possibility, which creates the volatility smile.

tively high, the option is considered expensive. when the implied volatility is relatively high and is excepted to drop, the best time to sell options.

Considerable research has been done to address the weaknesses of the Black-Scholes. Since Black-Scholes is widely used worldwide banks, it was the most popular to just model implied volatility and then keeping the Black-Scholes analytical formula as a tool to cite option prices from implied volatilities. standpoint of Riccardo Rebonato's Implied volatility is "the wrong number put in the wrong formula to get the right price", see [79]

To overcome the smile- effect, the market has to be extended. One widely used approach is to consider that the volatility also follows a stochastic diffusion process :

$$\begin{cases} dS = \mu S dt + \sigma S dW_t \\ d\sigma = \lambda_t dt + \xi_t dW_t^{\nu} \end{cases} \tag{2.15}$$

here W_t^{ν} is a further Wiener-process; the parameter λ_t, ξ_t have to be chosen appropriate.

2.1.5 Extension of the Black-Scholes model for volatility estimate

The critical parameter in the Black-Scholes model is the volatility σ. Assuming σ constant and using PDE-BS often leads to poor predictions of the option's prices, however it is known that the standard Black-Scholes model does not generate the observed option for all possible strike prices and expiry dates for any single value of the volatility. The volatility Smile is the price paid for too great simplicity of Black-Scholes assumption.

Several forms of volatility have been proposed to improve the **BS**-model, these areas of research lead to two tracks of extension of the Black-Scholes model:

1. **Local volatility models :** the volatility is a function of time and of the spot price, i.e. $\sigma_t = \sigma(t, S_t)$. With suitable assumptions on the regularity and the behavior at infinity of the function σ. The price option $p(t, S_t)$ satisfies the final value problem of PDE-BS, in which σ varies with t and S.

Calibrating the model consists of tuning the function σ in such a way that the prices computed e.g. with the PDE coincide with the observed prices.
In the early 90's *Dupire* (1994) on the one hand, and *Derman* & *kani* (1994) on the other, developed a concept called *Local Volatility* where the volatility smile was retrieved from the option prices. These two approaches are briefly presented:

- ***The Dupire Approach*** [60] : Suppose the option price of a Black-Scholes model with local volatility σ is denoted by

$$V := V(S,t) = V(S,t;K,T,r,q,\sigma).$$

Given $S_* > 0$, $t_* > 0$, we introduce a new variable u to using this price as follows:

$$u := u(K,T;S_*,t_*,r,q,\sigma) = V(S_*,t_*;K,T,r,q,\sigma).$$

Then on can show that the function u satisfies the following system:

$$\mathcal{L}_{Dup}(u) := \frac{\partial u}{\partial T} - \frac{1}{2}\sigma^2(K,T)K^2\frac{\partial^2 u}{\partial K^2} + (r-q)K\frac{\partial u}{\partial K} + qu = 0 \qquad K \in (0,\infty), T \in (t_*,\infty) \tag{2.16}$$

Further details for Dupire Equation is presented in the next chapter. Now, we can solve (2.16) for the Volatility to obtain

$$\boxed{\sigma(K,T)^2 = 2\frac{\frac{\partial u}{\partial T}(K,T) + (r-q)K\frac{\partial u}{\partial K}(K,T) + qu(K,T)}{K^2\frac{\partial^2 u}{\partial K^2}(K,T)}} \tag{2.17}$$

This formula has the consequence that in order to compute the local volatility once a solution of the equation (2.16) has to be found. The equation (2.17) can be used to estimate the LVF $\sigma(S_t,t)$ if option prices for all strikes and maturities are available in the market. Unfortunately in reality option prices are available only for a discrete set of strikes and maturities.

Thus the calibration of the local volatility function becomes recovering the local volatility from a finite set of market prices. This type of problem is ill-posed inverse problem, and it is the aim of the first part in this book.

- ***The Derman Kani Approach*** [57] : The Derman & Kani approach is similar to the above approach, except it uses the Binomial (or Trinomial) Tree framework instead of the continuous one.
Using the Binomial Tree notations, their upward transition probability p_i from the spot s_i at time t_n to the upper node S_{i+1} at the following time-step t_{n+1},is obtained from the usual

$$p_i = \frac{F_i - S_i}{S_{i+1} - S_i} \tag{2.18}$$

where F_i is the stock forward price known from the market, and S_i the lower spot at the step t_{n+1}. In addition, we have for a call expiring at time-step t_{n+1}

$$C(K, t_{n+1}) = e^{-r\Delta t} \sum_{j=1}^{n} [\lambda_i p_i + \lambda_{j+1}(1 - p_{j+1})] \max(S_{j+1} - K, 0)$$

where λ_j are the known Arrow-Derbou prices corresponding to the discounted probability of getting to the point s_j at time t_n from S_0 the initial stock price. We illustrate in the figure 2.2, the constructing level (n+1) of the implied binomial tree.

The implied local volatility $\sigma_{loc}(S_{n,i}, m\Delta t)$ that describes the structure of the second moment of the underlying process at any level m of the tree can then be calculated as a discrete approximation of the following conditional variance

$$\sigma_{loc}(S, m\Delta t) = Var\left(\log S_{t+\tau} | S_t = S\right),$$

where $S = S_{n,i}$ and $\tau = m\Delta t$

2. **Stochastic volatility models:** this type of models appeared in 1987, where modeling the dynamics with stochastic volatility was the subject of several publications.One assumes that $\sigma_t = f(y_t)$, where y_t is a continuous time stochastic process, correlated or not to the process driving S_t, (see [68] for a interest presentation). Serval such models have been propose, among which

   - ***Hull-White model*** (**1987**) [78]: considered $f(y) = \sqrt{y}$ and y_t is a lognormal process.

   - ***Stein-Stein*** (**1991**) [117]: considered $f(y) = \sqrt{y}$ and y_t is a is a mean reverting *Ornstein-Uhlenbeck*[2] process.

   - ***Heston model*** (**1993**) [76]: used $f(y) = \sqrt{y}$ where y_t *Cox-Ingersoll-Ross* (CIR) process,

     $$dy_t = \kappa(m - y_t)dt + \nu\sqrt{y_t}dW_t \qquad (2.19)$$

     where (W_t) a standard brownian motion and (y_t) is a process of the CIR type (Cox-Ingersoll-Ross), with a long-term average m and a return rate of average

[2]The Ornstein-Uhlenbeck process (named Leonard Ornstein and George Eugene Uhlenbeck), is a stochastic process that describes the velocity of a massive Brownian particle under the influence of friction. The process is stationary, Gaussian, and Markovian, up to allowing linear transformations of the space and time variables.

κ. The constant ν represents the diffusion of volatility.

- ***Scott model*** **(1997)** [115]: considered $f(y) = \sqrt{y}$ and y_t is a mean reverting *Ornstein-Uhlenbeck* process.

$$dy_t = \alpha(m - y_t)dt + \beta dW_t \tag{2.20}$$

where α and β are positive constants, W_t is a Brownian motion.

- ***Lévy driven spot price*** ([51], [9]): Lévy processes are processes with stationary and independent increment which are continuous in probability. For a Lévy process X_τ on a filtered probability space with probability $\mathbb{P}^*$, the Lévy-Khinchine formula says that there exists a function $\chi : \mathbb{R} \longrightarrow \mathbb{C}$ such that $\mathbb{E}(e^{iuX_\tau}) = e^{-\tau\chi(u)}$

$$\chi(u) = \frac{\sigma^2 u^2}{2} - i\beta u + \int_{|z|<1} (e^{iuz} - 1 - iuz)\nu(dz) + \int_{|z|>1} (e^{iuz} - 1)\nu(dz),$$

for $\sigma > 0$, $\beta \in \mathbb{R}$ and a positive measure ν on $\mathbb{R}\backslash\{0\}$ such that $\int_{\mathbb{R}} \min(1, Z^2)\nu(dz) < +\infty$. The measure ν is called the Lévy measure of X. And for the exponential of a Lévy process, the discounted price of the risky asset is a square integrable martingale under $\mathbb{P}^*$ is assumed. The pricing function for a European option is obtained by solving a partial integrodifferential equation (PIDE), with a non local term.

3. **NGARCH process :** The GARCH process called $GARCH(1,1)$ was developed originally in the field of econometrics in a discrete framework. The stock discrete equation (2.2) can be rewritten by taking $\Delta = 1$ and $v_n = \sigma_n^2$ as

$$\ln S_{n+1} = lnS_n + (\mu - \frac{1}{2}v_{n+1}) + \sqrt{(}v_{n+1})W_{n+1} \tag{2.21}$$

Calling the mean adjusted

$$u_n = \frac{\ln S_n}{lnS_{n-1}} - (\mu - \frac{1}{2}v_n) = \sqrt{(}v_n)W_n \tag{2.22}$$

the variance process in $GARCH(1,1)$ is supposed to be

$$v_{n+1} = \omega_0 + \beta v_n + \alpha u_n^2 = \omega_0 + \beta v_n + \alpha v_n W_n^2 \tag{2.23}$$

where α and β are weight parameters and ω_0 a parameter related to the long-term variance.

The limitation of the $GARCH(1,1)$ model, hence the introduction of the Non-Linear Asymmetric GARCH (NGARCH) model.

Duan[59] attempts to explain the volatility smile by using the NGARCH process expressed by

$$v_{n+1} = \omega_0 + \beta v_n + \alpha(u_n - c\sqrt{v_n})^2 \tag{2.24}$$

where c is a parameter to be determined.

2.1.6 Calibration of Local Volatility Function

The local volatility model are very elegant and theoretically sound, however they present in practice many stability issues. They are *Ill-Posed Inversion* problems and are extremely sensitive to the input data.

2.1.6.1 Optimization approach

To reconstruct the unknown volatility, it is a common procedure to transform the inverse problem in an optimization problem. Let the local volatility be parameterized in the following way:

$$\Sigma := \sigma(a)/a \in \mathcal{A}$$

where $\mathcal{A}$ is a set admissible parameters. The problem of calibration becomes the following optimization problem:

$$\min_{a\in\mathcal{A}} J :\mapsto \sum_{i=1}^{N}(V_{theo}(K_i, T_i)(a) - V_{mrkt}(K_i))^2 \tag{2.25}$$

Here $V_{theo}(.,.)$ are prices of the options derived from the Black-Scholes model and $V_{mrkt}(.)$ are the market prices. The set $\mathcal{A}$ where the volatility is to be found must be chosen in order to ensure that from minimizing sequence on can extract at least a subsequence that converges in $\mathcal{A}$ and that is limit is indeed a solution of the least square problem. We suppose for example, $\mathcal{A}$ a compact subset of a Hilbert space W such that the mapping J is continuous in W.

In practice, W has a finite dimension and is compactly embedded in the space of bounded and continuous functions σ. Thus, the existence of a solution to the minimization problem is most often guaranteed. *What is more difficult to guarantee is uniqueness and stability: is there a unique solution to the least square problem* ? If yes, *is the solution insensitive to small variation of data* ?
The answer to these question is that the problem is known has *ill-posed inverse problem* in the sense of Hadamard [73]).

2.1.6.2 Inverse problem

Modern option pricing theory came into existence with the advent of the influential paper by Black and Scholes [22]. The corresponding methods involves the solution of the Black-Scholes partial differential equation, which in turn depends on the risk-free interest rate prevalent in the market, the dividend rate, and the volatility of the underlying asset. There are many models describing the Local volatility as solution of the inverse problem, where σ is function depending on time t and underlying S. We can see the work of the authors *Dupire* [60], *Derman and Kani* [57], *Rubinstein* [113] and *Lagnado and Osher* [93].

In our work we will use the Dupire's equation in order to modelize and to solve our inverse problem. As usual, the dividend and interest rates are known during the option life. If the volatility σ is known, the Dupire boundary value problem is well posed in the sense of *Hadamard* [73] and allows the stable computation of the option price. Nevertheless, the volatility is not known explicitly, i.e. the unknown parameter in the initial value determined PDE of pricing is the volatility. Then, the inverse problem, can be solved if we have enough information on the prices of the option. But in practice:

- The inverse problem is underdetermined because a limited quantity of information provided in the market.
- The problem is ill-posed because the volatility is not continuously dependent on the market data.

The ill-posed character of the calibration local volatility function requires the use of regularization techniques. The famous model is the Tikhonov regularization [120] applied to a widely inverse problem in mathematical finance, that of calibrating a local volatility function from a given set of option prices, in a generalized Black-Scholes model.

This calibration problem has received intensive study in the last ten years, see for instance [26, 50, 54, 55, 60, 57, 93] and references therein. Notable approaches include entropy regularization (*Avellaneda et al.* [14]) or parametric expansion (*Bouchouev* and *Isakov*) [26]. In this first part of book, we shall focus upon the Tikhonov regularization method, following an approach introduced by *Lagnado* and *Osher* [93].
Then, The minimization function problem of (2.25) is modified, minimizing the functional $\sigma \mapsto J(\sigma) + J_R(\sigma)$ instead of J, where J_R is a sufficiently a strongly convex functional defined on W and containing some financially relevant information.

2.1.7 Local Volatility function for European Option

In this section, we review some facts about volatility calibration for the standard Black-Scholes model (see [22]) and Dupire model (see [60]) for pricing options. The

BS model assumes that the underlying asset price S satisfies the stochastic differential equation (2.1). Under non-arbitrage principle, the European option price V satisfies the parabolic partial differential equation.

Volatility calibration has received intense attention in the last decades. The list of references on the subject is too vast. For some interesting and recent results see [14, 26, 54, 60, 62, 93]. For example in [70], *Glover* used a set of radial basis functions, specially thin plate spline. in[124], Windcliff used Methods for grid construction and interpolation of jump conditions. And in [25], Bonnans used the parametrization cubic spline for calibration volatility and used Dupire PDE for direct problem.

In [93] Lagnado Osher proposed a regularized minimization method to fit the option prices using finite difference methods. The minimization process implies a regularization with restriction of σ to the smoothest functional that minimizes the difference between Black-Scholes prices and known market prices. Avellaneda et al. [14] proposed a representation of the volatility $\sigma(S,t)$ using a relative-entropy minimization method. It can generate volatilities that change suddenly and drawback numerical solution. A spline interpolation used in order to choose a smooth volatility function $\sigma(S,t)$ is presented in the work of Jackson et al. In [81]. This proposition is done via a regularization method that consists in a weighted minimization of the difference between the BS prices and the known market prices over a set of strikes and maturities. In [93], a penalization on the smoothness of $\sigma(S,t)$ is added to the minimization functional, this incorporation implies numerical stability and smoothness of the recovered volatility surface $\sigma(S,t)$. and the last example is the the model of C. Chiarella et al. in [34]. They developed an alternative strategy based on the work of Lagnado and Osher; to propose a fast algorithm for solving Computation of the volatility function. In our work [7], we are interested in the computation of the volatility σ, using an inverse problem, we will use modified Tikhonov regularization method.

2.2 Environment

In this part, we model saltwater intrusion for two aquifer configurations: confined and unconfined (free). We follow the derivation in [36] and adopt a mixed sharp/diffuse interface approach. We first recall the main features of porous media, then detail the derivation steps for each aquifer type, and finally present the mathematical results obtained (existence and uniqueness) for the different settings.

2.2.1 Basics of porous media

A porous medium can be viewed as a *solid with voids*. Pores are interconnected, forming continuous pathways from one side to the other, with a distribution that may be regular or random at the domain scale [17]. The solid phase is the *matrix*; the non-solid space is the *pore space*, occupied by liquid and/or gas. Porous media can be *natural* (rocks, soils) or *artificial* (concrete, cements).

2.2.1.1 Properties

Matrix geometry may be: **unconsolidated** (grains/fibers not bonded: gravels, sands, silts, glass/steel beads); or **consolidated** (compact matrix: limestone, sandstone, clays, wood, ceramics, ...).
Porosity. Void-to-total volume ratio (unitless). Total porosity

$$n = \frac{\text{void volume}}{\text{sample volume}},$$

and *effective* porosity

$$n_e = \frac{\text{gravitational water volume}}{\text{sample volume}}.$$

Permeability (k). Ability to transmit fluid under a pressure gradient (e.g., $1D \simeq 0.97 \times 10^{-12}\,m^2$).
Density (ρ). Mass per unit volume ($\mathrm{kg\,m^{-3}}$).
Relative density. Ratio of ρ to a reference density (pure water at 4°C for liquids/solids), indicating whether a substance floats or sinks in another.
Viscosity. Internal resistance to flow; higher viscosity implies more difficult and dissipative flow. For liquids, viscosity generally decreases with temperature. Dynamic viscosity μ obeys

$$F = \frac{\mu S\, dv}{dz},$$

and the kinematic viscosity is μ/ρ.
Piezometric head.

$$h_A = \frac{P_A}{\rho g} + z_A.$$

Hydraulic head.

$$\Phi_A = h_A + \frac{v^2}{2g},$$

with the kinetic term negligible in porous media.
Hydraulic conductivity.

$$K = \frac{k\,\rho g}{\mu} \quad (\mathrm{m\,s^{-1}}).$$

2.2.1.2 Fluid types

Compressible/incompressible: gases are compressible; liquids are modeled as incompressible.
Newtonian/non-Newtonian: Newtonian fluids have stress-independent viscosity (e.g., water); non-Newtonian fluids do not.

2.2.2 Aquifer classification

An aquifer is a permeable formation holding significant water volumes:

- **Unconfined (water table) aquifer:** permeable top, free piezometric surface, exchanges with the exterior (recharge, pumping).
- **Confined aquifer:** bounded by two impermeable layers; groundwater is under pressure and may rise above the aquifer top in wells.
- **Semi-confined:** semi-impermeable top/bottom allowing slow vertical exchanges.

Subsurface is divided into an **unsaturated zone** (air + water in pores, negative water pressure) and a **saturated zone** (pores fully water-filled, positive pressure).

2.2.3 Model derivation

We follow [36].

2.2.3.1 Governing equations

Continuity (mass conservation). With sources,

$$\frac{\partial(\rho\phi)}{\partial t} + \nabla \cdot (\rho q) = \rho Q. \qquad (??)$$

Darcy's law. For laminar saturated flow,

$$q = -\frac{k}{\mu}(\nabla P + \rho g \nabla z_h),$$

or, in hydraulic head variables,

$$q = -K\nabla\Phi + \frac{k}{\mu}(\rho - \rho_0) g \nabla z_h. \qquad (??)$$

Constitutive (state) relations. Elastic model (water/matrix):

$$\frac{d\rho}{\rho} = \beta_P \, dP, \qquad \frac{dV}{V} = -\alpha_P \, d\bar{\sigma} = \alpha_P \, dP,$$

with Terzaghi's relation $dP + d\bar{\sigma} = 0$ and $V_s = (1-\phi)V$, yielding

$$d(\rho\phi) = \rho\big(\phi\beta_P + (1-\phi)\alpha_P\big)\, dP.$$

2.2.3.2 Simplifying assumptions

Low compressibility. Liquids and soil are slightly compressible ($\beta_P, \alpha_P \ll 1$), hence $\rho \simeq \rho_0$ and

$$q = -K\nabla\Phi, \qquad K = \frac{k\rho_0 g}{\mu}. \tag{??}$$

Bear's hypothesis. Spatial variations of ρ are small compared to temporal ones [17, 19]:

$$\nabla \cdot (\rho q) \approx \rho \nabla \cdot q,$$

so

$$\rho_0 g\big(\phi\beta_P + (1-\phi)\alpha_P\big)\,\partial_t\Phi + \nabla \cdot q = Q, \tag{??}$$

with storage $S_0 = \rho_0 g\big((1-\phi)\alpha_P + \phi\beta_P\big)$.
Sharp interface. The domain is split into freshwater/saltwater (Fig. ??). In each subdomain:

$$S_f\partial_t\Phi_f + \nabla \cdot q_f = Q_f, \qquad S_s\partial_t\Phi_s + \nabla \cdot q_s = Q_s, \tag{??–??}$$

and $q_\bullet = -K_\bullet\nabla\Phi_\bullet$.
Dupuit approximation. Small slopes $\Rightarrow$ quasi-horizontal flow ($q_z \ll 1$), vertical integration and $3D \to 2D$ reduction.

2.2.3.3 Integration domain and conditions

With $\Omega \subset \mathbb{R}^2$, bottom h_2, interface h, and top h_1, the $3D$ systems, dry/river boundary zones, and vertical integration (Leibniz) lead to

$$S_fB_f\partial_t\tilde{\Phi}_f - \nabla_{xy}\cdot(B_f\tilde{K}_{f_{xy}}\nabla_{xy}\tilde{\Phi}_f) + q_f(h_1^-)\cdot\nabla(z-h_1^-) - q_f(h^-)\cdot\nabla(z-h^-) = B_f\tilde{Q}_f. \tag{2.26}$$

and,

$$S_sB_s\partial_t\tilde{\Phi}_s - \nabla_{xy}\cdot(B_s\tilde{K}_{s_{xy}}\nabla_{xy}\tilde{\Phi}_s) + q_s(h^-)\cdot\nabla(z-h^-) - q_s(h_2)\cdot\nabla(z-h_2) = B_s\tilde{Q}_s, \tag{2.27}$$

with $B_f = h_1^- - h^-$ and $B_s = h^- - h_2$.

2.2.3.4 Continuities and interface fluxes

Constant viscosity. Homogeneous/isotropic medium and $\mu_f = \mu_s$ imply $\tilde{K}_s = (1+\alpha)\tilde{K}_f$, with $\alpha = \rho_s/\rho_f - 1$.
Continuous pressure at $z = h$. Under Dupuit:

$$(1+\alpha)\nabla\tilde{\Phi}_s = \nabla h_1^- + \alpha\nabla h^-, \quad (1+\alpha)\partial_t\tilde{\Phi}_s = \partial_t h_1^- + \alpha\partial_t h^-, \tag{??–??}$$

and Allen?Cahn phase-field flux conditions yield

$$q_f(h^-)\cdot\nabla(z-h^-) = q_s(h^-)\cdot\nabla(z-h^-) = \phi\Big(\partial_t h^- - \nabla\cdot(\delta_h\nabla h^-)\Big), \tag{??}$$

$$q_f(h_1^-)\cdot\nabla(z-h_1^-)=\begin{cases}0, & \text{confined,}\\ \phi\Big(\partial_t h_1^- - \nabla\cdot(\delta\nabla h_1^-)\Big), & \text{unconfined,}\end{cases} \tag{??,??}$$

$$q_s(h_2)\cdot\nabla(z-h_2)=0. \tag{??}$$

2.2.4 Models by aquifer type

Neglecting $S_s B_s \partial_t \tilde{\Phi}_s$ (saltwater nearly stationary and slightly compressible), combining the integrated equations, continuity relations, and flux laws yields reduced systems.

2.2.4.1 Confined aquifer

With normalizations ($\delta_h = \delta$, $\nabla_{xy}=\nabla$, $\tilde{K}_f=K$, $\tilde{\Phi}_f=\Phi_f$) and vertical axis inversion, the unknowns are h (fresh/salt interface) and Φ_f (freshwater head). The final system is

$$(\mathcal{M}_1)\quad\begin{cases}\phi\,\partial_t h+\nabla\cdot\big(K(h_2-h)\nabla\Phi_f\big)-\nabla\cdot\big(\alpha K(h_2-h)\nabla h\big)-\phi\,\nabla\cdot(\delta\nabla h)=(h-h_2)\,\tilde{Q}_s,\\ -\nabla\cdot\big(K(h_2-h_1)\nabla\Phi_f\big)+\nabla\cdot\big(\alpha K(h_2-h)\nabla h\big)=(h-h_1)\,\tilde{Q}_f+(h_2-h)\,\tilde{Q}_s.\end{cases} \tag{??}$$

2.2.4.2 Unconfined aquifer

The unknowns are h (fresh/salt interface) and h_1 (saturated/unsaturated interface). The final system is

$$(\mathcal{M}_2)\quad\begin{cases}\phi\,\partial_t h-\nabla\cdot\big(K(h_2-h)\nabla h_1\big)-\nabla\cdot\big(\alpha K(h_2-h)\nabla h\big)-\phi\,\nabla\cdot(\delta\nabla h)=(h-h_2)\,\tilde{Q}_s,\\ \big[S_f(h-h_1)+\phi\big]\partial_t h_1-\nabla\cdot\big(K(h_2-h_1)\nabla h_1\big)-\nabla\cdot\big(\alpha K(h_2-h)\nabla h\big)\\ \qquad\qquad -\phi\,\nabla\cdot(\delta\nabla h_1)=(h_1-h)\,\tilde{Q}_f+(h-h_2)\,\tilde{Q}_s.\end{cases} \tag{??}$$

CHAPTER 3

Calibration of Volatility: Coupling between alternative regularization strategy and Dupire's Equation

Collaborated with : **R. Aboulaich**[1]

Contents

[1]EMI,Mohammed V University, Rabat

We propose in this chapter a modified Tikhonov regularization method for a calibration problem. The goal is to identify the local volatility from some observed option prices on the market. The proposed method, is used by Chiarella et al. [34], Lagnado and Osher [93] using a Black-Scholes model, in our work we propose to use Dupire model.

In order to improve the approximation of the initial guess of the volatility we propose to use a statistical approximation. The obtained results are compared with those obtained by Chiarella et al. in [34]. and with Aboulaich et al. [6] which proposes another algorithm named "jump indicator" in the case of a discontinuous volatility.

3.1 Direct Problem: Dupire Partial Differential Equation

In this section we consider the problem of inferring an effective volatility function $\sigma(S,t)$ from observed option prices, namely Vanilla calls and/or puts with various strikes and maturities on the underlying S.

3.1.1 Dupire Partial Differential Equation

One way to make model prices consistent with market prices for European Option is to assume that the volatility is a function of time and the underlying price. The price of the asset S satisfies the differential equation given in [25]:

$$dS = S(r-q)dt + S\sigma(S,t)dW \tag{3.1}$$

The theorem of Kolmogorov is given in [90] as follows:

Theorem 1.

For an Itô process given by the stochastic differential equation (SDE):

$$dX_t = \mu(X_t,t)dt + X_t\sigma(t,X_t)dW_t, \quad X_0 = x$$

with drift $\mu(X_t,t)$ and diffusion coefficient $X_t\sigma(t,X_t)$, and Winer process W_t. Then the probability density $\phi(x,t)$ of the random variable X_t satisfies the PDE

$$\frac{\partial}{\partial t}\phi(x,t) = -\frac{\partial}{\partial x}(\mu(x,t)\phi(x,t)) + \frac{1}{2}\frac{\partial^2}{\partial x^2}(x^2\sigma^2(x,t)\phi(x,t))$$

The Kolmogorov forward equation is sometimes also called Fokker-Plank's equation, (we note $X_t = S_t$, $\mu(X_t, t) = S(r-q)$), the equation is given by:

$$\partial_t p(S,t) = -\partial_S((r-q)Sp(S,t)) + \partial_{SS}(\frac{\sigma(S,t)^2}{2}S^2 p(S,t)), \tag{3.2}$$

Where $p(S,t)$ is the density of the underlying asset price in t .
The option's price with a Strike K and maturity T in t_0 is given by :

$$V(K,T) = \exp(-r(T-t_0))\int_0^\infty (S-K)_+ p(S,t)dS \tag{3.3}$$

We derive with respect t, and using equation (3.2) we obtain:

$$\partial_t V(K,T) = -rV(K,T) + \exp(-r(T-t0)))\int_K^\infty (S-K)_+[\partial_S((r-q)Sp(S,t)) + \partial_{SS}(\frac{\sigma(S,t)^2}{2}S^2 p(S,t))]dS \tag{3.4}$$

Using an integration by parts, we obtain:

$$\partial_t V(K,T) = -rV(K,T) + e^{-r(T-t_0)}\frac{1}{2}\sigma^2(K,T)K^2 p(K,T) \tag{3.5}$$

$$+ (r-q)e^{-r(T-t_0)}\int_K^\infty (S-K)_+ p(S,T)dS + K(r-q)e^{-r(T-t_0)}\int_K^\infty p(S,T)dS \tag{3.6}$$

Observe that:

$$V(K,T) = e^{-r(T-t_0)}\int_K^\infty (S-K)_+ p(S,T)dS$$
$$\partial_K V(K,T) = e^{-r(T-t_0)}\int_K^\infty p(S,T)dS$$
$$\partial_{KK} V(K,T) = e^{-r(T-t_0)} p(K,T)$$

Then, the new equation is given by:

$$\frac{\partial V}{\partial T}(K,T) = -qV(K,T) - (r-q)K\frac{\partial V}{\partial K}(K,T) + \frac{1}{2}\sigma^2(K,T)K^2\frac{\partial^2 V}{\partial K^2}(K,T) \tag{3.7}$$

The boundary conditions are given by:

$$\begin{cases} U(K,t_0) = max(S_0 - K, 0) & \text{for} \quad 0 \le y \le K \\ \lim_{K\to 0} V(K,T) = S_0 e^{-q(T-t_0)} & \text{for} \quad T \ge t_0 \\ \lim_{K\to +\infty} V(K,T) = 0 & \text{for} \quad T \ge t_0. \end{cases} \tag{3.8}$$

3.1.2 Transformation into logarithmic variables

To obtain a less restrictive stability condition on time and space steps, we choose to use a logarithmic variables. We set:

$$y = \ln(K),$$
$$U(y,T) = V(K,T),$$
$$\widehat{\sigma}(y,T) = \sigma(K,T).$$

We obtain the Following PDE:

$$\frac{\partial U}{\partial T}(y,T) = -qU(y,T) - (r - q + \frac{1}{2}\widehat{\sigma}^2(y,T))\frac{\partial U}{\partial y}(y,T) + \frac{1}{2}\sigma^2(y,T)\frac{\partial^2 U}{\partial y^2}(y,T) \tag{3.9}$$

We rewrite the boundary conditions using a logarithmic variables. Then, the new conditions are given as follows :

$$\begin{cases} U(y,t_0) = max(S_0 - e^y, 0) & \text{for} \quad y_{min} \leq y \leq y_{max} \\ \lim_{y\to-\infty} U(y,T) = S_0 e^{-q(T-t_0)} & \text{for} \quad T \geq t_0 \\ \lim_{y\to+\infty} U(y,T) = 0 & \text{for} \quad T \geq t_0. \end{cases} \tag{3.10}$$

3.1.3 Discretization

3.1.3.1 Space-discretization

We defined the operator $\widetilde{A}$ as follows:

$$\widetilde{A} = -\frac{1}{2}\sigma^2(y,T)\frac{\partial^2}{\partial y^2} - q + (r - q + \frac{1}{2}\widehat{\sigma}^2(y,T))\frac{\partial}{\partial y}$$

And we note:

$$(\widetilde{A}U)(y,T) = -\frac{1}{2}\sigma^2(y,T)\frac{\partial^2 U}{\partial y^2}(y,T) - qU(y,T) + (r - q + \frac{1}{2}\widehat{\sigma}^2(y,T))\frac{\partial U}{\partial y}(y,T)$$

Then, the equation (3.9) is written as follow:

$$\frac{\partial U}{\partial T}(y,T) + (\widetilde{A}U)(y,T) = 0 \tag{3.11}$$

For all $y \in [y_{min}, y_{max}]$ and $T \in [t_0, T_{max}]$ we have:

$$\begin{cases} \frac{\partial U}{\partial T}(y,T) + (\widetilde{A}U)(y,T) = 0 & y_{min} \leq y \leq y_{max}, t_0 \leq \tau \leq T_{max} \\ U(y,t_0) = max(S_0 e^y, 0) & for \quad y_{min} \leq y \leq y_{max} \\ \lim_{y\to-\infty} U(y,\tau) = S_0 e^{-q(T-t_0)} & for \quad t_0 \leq T \leq T_{max} \\ \lim_{y\to+\infty} U(y,T) = 0 & for \quad t_0 \leq T \leq T_{max} \end{cases} \tag{3.12}$$

We note:
$h = (y_{max} - y_{min})/N$ step space,
$y_i = y_{min} + i * h,$
$f_i = f(y_i)$

Considering $U(T) = (U(y_i, T))_{1 \leq i \leq N-1} \in \mathbb{R}^{N-1}$
For all $T \in [t_0, T_{max}]$, discretizing $\widetilde{A}$ for the discrete operator
$\widetilde{A}_T : U(T) \in \mathbb{R}^{N-1} \to \widetilde{A}_T U(T) \in \mathbb{R}^{N-1}$, we have :

$$\begin{cases} \frac{\partial U}{\partial y}(y_i, T) \simeq \frac{U_{i+1}(T) - U_{i-1}(T)}{2h}, \\ \frac{\partial^2 U}{\partial y^2}(y_i, T) \simeq \frac{U_{i+1}(T) - 2U - i(T) + U_{i-1}(T)}{h^2}. \end{cases}$$

we note:

$$\begin{cases} \alpha_{i,T} = -\frac{\widehat{\sigma}^2(y_i,T)}{2h^2} - \frac{1}{2h}(r - q + \frac{1}{2}\widehat{\sigma}^2(y_i, T)) \\ \beta_{i,T} = \frac{\widehat{\sigma}^2(y_i,T)}{h^2} + q \\ \gamma_{i,T} = -\frac{\widehat{\sigma}^2(y_i,T)}{2h^2} \frac{1}{2h}(r - q + \frac{1}{2}\widehat{\sigma}^2(y_i, T)) \end{cases}$$

Therefore, for $i \in 1, ..., N-1$ et for all T:

$$(\widetilde{A}_T U(T))_i = \alpha_{i,T} U_{i-1}(T) + \beta_{i,T} U_i(T) + \gamma_{i,T} U_{i+1}(T) \tag{3.13}$$

For the boundary conditions:

<u>$For \quad i = 0,$</u> we have the Dirichlet condition for T : $U_0(T) = S_0 e^{-q(T-t_0)}$ Then,

$$(\widetilde{A}_T U(T))_1 = \alpha_{1,T} S_0 e^{-q(T-t_0)} + \beta_{1,T} U_1(T) + \gamma_{1,T} U_2(T) \tag{3.14}$$

<u>$For \quad i = N,$</u> $(y = y_{max})$ we have : $U_N(T) = 0$ Then,

$$(\widetilde{A}_T U(T))_{N-1} = \alpha_{N-1,T} U_{N-2}(T) + \beta_{N-1,T} U_{N-1}(T). \tag{3.15}$$

<u>$For \quad T = t_0,$</u> we have : $f_0 = Se^{-q(T-t_0)}, \quad f_N = 0$ Then,

$$\begin{cases} (\widetilde{A}_T U(T))_1 = \alpha_{1,T} S_0 \exp^{-q(T-t_0)} + (\widehat{A}_T U(T))_1 \\ (\widetilde{A}_T U(T))_i = (\widehat{A}_T U(T))_i, \quad \forall \quad i \in 2, ..., N-1 \end{cases} \tag{3.16}$$

where,

$$(\widehat{A}_T)_{1 \leq i,j \leq N-1} = \begin{pmatrix} \beta_{1,T} & \gamma_{1,T} & 0 & \cdots & 0 \\ \alpha_{2,T} & \beta_{2,T} & \gamma_{2,T} & \cdots & 0 \\ 0 & \ddots & \ddots & \ddots & \vdots \\ \vdots & & \alpha_{N-2,T} & \beta_{N-2,T} & \gamma_{N-2,T} \\ 0 & \cdots & \cdots & \alpha_{N-1,T} & \beta_{N-1,T} \end{pmatrix}$$

Finally, the spatial discretization for the equation (3.9) and is given:

$$(E^h): \begin{cases} \frac{\partial U}{\partial T}(T) + \widehat{A}_T U(T) + \alpha_{1,T} S_0 \exp^{-q(T-t_0)} .e_1 = 0, \quad for \quad T \in [t_0, T_{max}] \\ U(t_0) = f \end{cases} \tag{3.17}$$

where, $f = (f(y_i))_{1 \le i \le N-1}$ and $e_1 = (1, 0, \cdots, 0)^T$

3.1.3.2 Time discretization: Θ-schema

we note $k = \frac{T_{max}-t_0}{M}$ step temporal

The solution U of the system (E^h) of 4.21 is written as follow:

$$(E^{h,k}): \begin{cases} \frac{u^{n+1}-u^n}{k} + \Theta(\widehat{A}_T^n U^n + \alpha_{1,T_n} S_0 \exp^{-q(T_n-t_0)} .e_1) \\ +(1-\Theta)(\widehat{A}_T^{n+1} U^{n+1} + \alpha_{1,T_{n+1}} S_0 \exp^{-q(T_{n+1}-t_0)} .e_1) = 0, \quad for \quad 0 \le n \le M-1 \\ U^0 = f \end{cases} \tag{3.18}$$

The Euler-Schema ($\Theta = 1$)

$$(E^{h,k}): \begin{cases} \frac{u^{n+1}-u^n}{k} + \widehat{A}_T^n U^n + \alpha_{1,T_n} S_0 \exp^{-q(T_n-t_0)} .e_1 = 0, \quad for \quad 0 \le n \le M-1 \\ U^0 = f \end{cases} \tag{3.19}$$

$\Rightarrow$

$$\boxed{U^{n+1} = U^n - k(\widehat{A}_T^n U^n + \alpha_{1,T_n} S_0 \exp^{-q(T_n-t_0)} .e_1)} \tag{3.20}$$

3.2 Regularization problem

3.2.1 Ill-posed Inverse problem

The origin of the inverse problem is mirrors what is called the direct problem. In simple terms, the direct problem is **"given question, find answer"**, whereas the inverse problem is **" given answer, find question"**. The simple example is given in [88], *find a polynomial p of degree n that assumes given values* $y_1, ..., y_n \in \mathbb{R}$ *at given points* $x_1, ..., x_n \in \mathbb{R}$. *This problem is inverse to the direct problem of calculating the given polynomial at given* $x_1, ..., x_n$. *The inverse problem is the Lagrange interpolation problem.* In other words, the solution for inverse problem entails determining unknown causes bases on observation of their effects.

The formulation of the direct problem is as the evaluation of an operator K, acting on a known "model" x in a model space X and the inverse problem as the solution of the equation $K(x) = y$:

- Direct problem: given x (and K), evaluate $K(x)$.

- Inverse problem: given y (and K), solve $K(x) = y$ for x.

In the last decades, the option price inverse problem has attracted a large amount of research and attention. The idea is, given a few assumptions about the underlying process, to attempt to derive the corresponding dynamics in the risk neutral measure from the option price observation. A correct derivation of the underlying process requires a complete knowledge of the volatility.

The problem of calibration Volatility is ill-posed problem because the set of price observations is discrete and finite. To understand what ill-posed problem is suffices to know the definition of Hadamard for well-posed problem. The general theory of inverse problem is discussed in [88] and [34].

Definition 2. *(* ***well-posedness****, given in [88])*
Let X and Y be normed spaces, $K : X \mapsto Y$ a (linear or nonlinear) mapping. The equation $Kx = y$ is called properly-posed or well-posed if the following holds:

1. ***Existence:*** *For every $y \in Y$ there is (at least one) $x \in X$ such that $Kx = y$.*

2. ***Uniqueness:*** *For every $y \in Y$ there is at most one $x \in X$ with $Kx = y$*

3. ***Stability:*** *The solution x depends continuously on the data y.*

Equations for which (at least) one of these properties does not hold are called ill-posed.

In our case y is the price of the option and is denoted by V, and x is the volatility and is denoted by y. The volatility function σ can not be uniquely determined with guaranteed continuous dependance on the market price observations, ie. small perturbation in the price data can lead to large changes in the minimizing function.

In order to deal with the ill-posedness of the calibration problem, regularization is usually required. There are many approaches to constructing regularization strategies, Tikhonov regularization is the most popular one. It's will be discussed in the next section.

3.2.2 Tikhonov regularization

The problem inverse proposed is ill posed, one tries to ameliorate this problem by introducing a regularization strategy given in [88].

Definition 3. *A regularization strategy for the equation $Kx = y$ is a family of linear and bounded operators*

$$R_\alpha : Y \to X \,,\ \alpha > 0 \textit{ such that}$$
$$lim_{\alpha \to 0}\ R_\alpha\ Kx = x,\ \forall\ x \in X$$

i.e. the operators $R_\alpha K$ converge pointwise to the identity. The real number α is known as the regularization parameter.

The definition of a regularization strategy assumes that we known the data y for the equation $Kx = y$ exactly. In practice however we have data y^ε where $||y - y^{epsilon}|| \leq \varepsilon$. If we apply the regularization strategy to the problem with perturbed data we obtain a family of approximation solutions

$$x^{\alpha,\varepsilon} = R_\alpha y^\varepsilon$$

to our real problem $Kx = y$.
The origin of the so called Tikhonov regularization scheme lies with the problem of dealing with an over determined linear system $Kx = y$, where K is a linear operator on a finite dimensional vector space. i.e. a matrix.
Tikhonov proposed the following alternative. Given the problem of solving $Kx = y$ where $K : X \to Y$ is a bounded linear operator, we instead determine the minimum value of the Tikhonov functional

$$J_\alpha(x) = ||Kx - y||^2 + \alpha||x||^2 \tag{3.21}$$

The positive real number α is known as the regularization parameter.

Theorem 2.

Let$K : X \to Y$ be a linear and bounded operator between Hilbert spaces and let $\alpha > 0$. Then the functional 3.21 has a unique minimum $x^\alpha \in X$, which is the unique solution of the normal equation

$$\alpha x^\alpha + K^* K x^\alpha = K^* y \tag{3.22}$$

Here the operator $K^* : Y \to X$ is the adjoint of the operator K. i.e the operator such that :

$$(Kx, y)_Y = (x, K^* y)_X$$

for all $x \in X$, $y \in Y$. $(.,.)_X$ and $(.,.)_Y$ are the inner products for the Hilbert spaces X and Y respectively

Proof. Let $(x_p)_{p\in\mathbb{N}} \subset X$ be a minimizing sequence; that is, $\lim_{p\to\infty} J_\alpha(x_p) = S := \inf_{x\in X} J_\alpha(x)$. For the existence of a minimum of J_α, we show that (x_p) is a cauchy sequence. Let $(p,q) \in \mathbb{N}^2$, we have:

$$\begin{aligned} J_\alpha(x_q) - J_\alpha(x_p) &= 2J_\alpha(\frac{1}{2}(x_q + x_p)) + \frac{1}{2}\|K(x_q - x_p)\|^2 + \alpha 2\|x_q - x_p\|^2 \\ &\geq 2S + \frac{\alpha}{2}\|x_q - x_p\|^2. \end{aligned}$$

The left-hand side converges to 2S when p, q tend to infinity. This shows that (x_p) is a Cauchy sequence and thus convergent. Then, let $x^\alpha = \lim_{p\to\infty} x_p$, noting that $x^\alpha \in X$. From the continuity of J_α, we conclude that $J_\alpha(x_p) \to J_\alpha(x^\alpha)$; that is, $J_\alpha(x^\alpha) = S$.
This proves the existence of a minimum of J_α. Using the lemma 2.10 introduce in the book of Kirsh [88] page 36, for the uniqueness of the solution of the normal equation. □

Tikhonov regularization is a compromise between precision and stability.
In the following we will use σ and V instead of x and y, σ denoting the volatility parameter and V the option value.
The Tikhonov regularization for the calibration problem of local volatility is writing as follow:

$$J(\sigma) = G(\sigma) + \alpha\|\sigma\|_2^2 \tag{3.23}$$

which is of course equivalent to minimizing

$$\lambda G(\sigma) + \|\sigma\|_2^2 \tag{3.24}$$

3.3 Alternative Regularization Strategy for Calibration of Local Volatility

3.3.1 Regularization Strategy for Calibration of Local Volatility

In the following, we describe briefly the approach proposed by [34]. In general context, market calibration involves finding a local volatility function σ such that solution of (3.1) fall between the corresponding bid and ask market,

$$V_{ij}^b \leq V(S_0, 0, K_{ij}, T_i, \sigma) \leq V_{ij}^a \tag{3.25}$$

To satisfy (3.25) we try to minimize the functional

$$G(\sigma) = \sum_{i=1}^{N}\sum_{j=1}^{M_i}[V(S_0, 0, K_{ij}, T_i, \sigma) - V_{ij}]^2 \tag{3.26}$$

Since the problem (3.26) is ill-posed,the function σ cannot be uniquely determined with guaranteed continuous dependence on the market price data. Therefore, this problem is reduced by a regularization technique. Lagnado and Osher [93] introduce of the regularization based upon the Tikhonov functional (see Tikhonov [120]).

Regularization strategy of Lagnado and osher[93] is based upon minimizing the functional:

$$\min F(\sigma) := \|\nabla\sigma\|_2^2 + \lambda G(\sigma) \tag{3.27}$$

Where G is given by (3.26), and $\lambda > 0$ is the parameter of regularization.
Let us rewrite the right hand side of(3.27) as an integral:

$$\int_0^\infty \int_0^\infty ((\frac{\partial\sigma}{\partial S})^2 + (\frac{\partial\sigma}{\partial t})^2 + \lambda\sum_{i=1}^{N}\sum_{j=1}^{M_i}[V(S,t,K_{ij},T_i,\sigma) - V_{ij}]^2 h(S,t))dSdt \tag{3.28}$$

where :

$$h(S,t) = \delta(S - S_0)\delta(t)$$

So that, we apply the Euler-Lagrange Equation from the calculus of variations to

$$L = (\frac{\partial\sigma}{\partial S})^2 + (\frac{\partial\sigma}{\partial t})^2 + \lambda\sum_{i=1}^{N}\sum_{j=1}^{M_i}[V(S,t,K_{ij},T_i,\sigma) - V_{ij}]^2 h(S,t) \tag{3.29}$$

and we obtain the Euler-Lagrange equation for the functional $F(\sigma)$

$$\frac{\partial^2\sigma}{\partial S^2} + \frac{\partial^2\sigma}{\partial t^2} - \lambda\sum_{i=1}^{N}\sum_{j=1}^{M_i}\frac{\delta V^{ij}}{\delta\sigma}[V(S_0,0,K_{ij},T_i,\sigma) - V_{ij}] = 0 \tag{3.30}$$

Thus to determine the function $\sigma(S,t)$ we must solve equation (3.30). Lagnado and Osher's regularization strategy calibrates the model for one fixed value of the index S_0, at one fixed point in the time $t = 0$. Consequently, there is no guarantee that the value σ determined by the procedure of Lagnado and Osher [93] will be correct for other value of index or for future times. The alternative presented by Chiarella et al. [34], take into account the variation of option price with changing value of the index over time. In the other words she attempt to minimize.

$$H(\sigma) = \sum_{i=0}^{N}\sum_{j=1}^{M_i}\int_0^\infty \int_0^{T_{cur}} (V(S,t,K_{ij},T_i,\sigma) - V^{ij})^2 dSdt \tag{3.31}$$

Where T_{cur} is the current time. This regularization strategy has the technique advantage that it contains no terms involving the Dirac Delta function. Consequently, the standard results from the calculus of variations can be applied.
The Euler-Lagrange equation for (3.31) is:

$$\frac{\partial^2\sigma}{\partial S^2} + \frac{\partial^2\sigma}{\partial t^2} - \lambda\sum_{i=1}^{N}\sum_{j=1}^{Mi}\frac{\partial V}{\partial\sigma}(S,t,K_{ij},T_i,\sigma)(V(S,t,K_{ij},T_i,\sigma) - V^{ij}) = 0 \tag{3.32}$$

Then we solve the equation given by :

$$\frac{\partial^2 \sigma}{\partial S^2} + \frac{\partial^2 \sigma}{\partial t^2} = \lambda W(S, t, K_{ij}, T_i, \sigma) \tag{3.33}$$

Where

$$W(S, t, K_{ij}, T_i, \sigma) = \sum_{i=1}^{N} \sum_{j=1}^{Mi} \frac{\partial V}{\partial \sigma}(S, t, K_{ij}, T_i, \sigma)(V(S, t, K_{ij}, T_i, \sigma) - V^{ij})$$

In order to determine the function $\sigma(S, t)$ we must to solve equation (3.33).

3.3.2 Alternative regularization strategies using Dupire's equation

In [34], the regularization strategy for the calibration of the volatility is used Black-Scholes equation for the computation of the theoretical price of an Option $V(S, t, K, T, \sigma)$, in this work we use the Dupire equation (See section 3.1).

In the Black-Scholes PDE, the parameters of the problem are the strike K and the maturity T of the option under consideration, and the variables are t and S. However, in the Dupire PDE, the point of view is different: (S_0, t_0) are the parameters of the problem and the variables are K and T. Therefore, when we solve the Dupire PDE, we get a layer of option prices on K and T for (S_0, t_0) fixed.

As introduced preciously in section 3.3.1, we will minimize the functional:

$$J(\sigma) = \|\sigma\|_2^2 + \lambda H_1(\sigma) \tag{3.34}$$

Where,

$$H_1(\sigma) = \sum_{i=1}^{N} \sum_{j=1}^{M} [U(y_j, T_i, \sigma) - U_{ij}]^2$$

and $y = \ln(K)$, $U(y, T, \sigma)$ is the price of an Option given by the Logarithmic Dupire's equation.

We inspired of the calculation made in Section 3.3, to minimize $J(\sigma)$. We apply the Euler-Lagrange equation for the function $L(\sigma)$, which is given as follows

$$L(\sigma) = (\frac{\partial \sigma}{\partial S})^2 + (\frac{\partial \sigma}{\partial t})^2 + \lambda H_1(\sigma) \tag{3.35}$$

In theorem below we present the Euler-Lagrange equation, which is given in[114].

Theorem 3.

Let $\Omega \in \mathbb{R}^d$ and $f \in C^2(\Omega)$ the extremum of the functional Φ given by:

$$\Phi[f] = \int_\Omega F(\underline{x}, f(\underline{x}), f'_{x_1}, (\underline{x}), ..., f'_{x_d}(\underline{x}))dx$$

with the boundary conditions $f(\underline{x}) = f_0(\underline{x})$ fixed. for all $\underline{x} \in \partial\Omega$ Then f necessarily satisfies the corresponding Euler-Lagrange equation

$$\frac{\partial F}{\partial f} - \sum_{i=1}^{d} \frac{\partial}{\partial x_i} \frac{\partial F}{\partial f'_{x_i}} = 0. \tag{3.36}$$

Now, we applied the theorem 2 for the minimization problem

$$F(\sigma) = \int_\Omega L(\sigma)d\Omega$$

Then,

$$\frac{\partial L}{\partial \sigma} - \sum_{i=1}^{2} \frac{\partial}{\partial x_i} \frac{\partial L}{\partial \sigma'_{x_i}} = 0$$

$\Rightarrow$

$$\frac{\partial L}{\partial \sigma} - \frac{\partial}{\partial S} \frac{\partial L}{\partial \sigma'_S} - \frac{\partial}{\partial t} \frac{\partial L}{\partial \sigma'_t} = 0$$

$$\frac{\partial}{\partial S} \frac{\partial L}{\partial \sigma'_S} + \frac{\partial}{\partial t} \frac{\partial L}{\partial \sigma'_t} - \frac{\partial L}{\partial \sigma} = 0 \tag{3.37}$$

According to equation (3.37), three terms must be calculated.

- $\frac{\partial}{\partial S} \frac{\partial L}{\partial \sigma'_S} = \frac{\partial}{\partial S} \frac{(\frac{\partial \sigma}{\partial S})^2 + (\frac{\partial \sigma}{\partial t})^2 + \lambda H_1(\sigma)}{\frac{\partial \sigma}{\partial S}}$

  The$\frac{\partial \sigma}{\partial t}^2$ and $H_1(\sigma)$ are constants with respect to $\frac{\partial \sigma}{\partial S}$, then we have:

  $$\frac{\partial}{\partial S} \frac{\partial L}{\partial \sigma'_S} = \frac{\partial}{\partial S} (\frac{\partial \sigma}{\partial S})^2 \Rightarrow \boxed{\frac{\partial}{\partial S} \frac{\partial L}{\partial \sigma'_S} = 2\frac{\partial^2 \sigma}{\partial S^2}.}$$

- $\frac{\partial}{\partial t} \frac{\partial L}{\partial \sigma'_t} = \frac{\partial}{\partial t} \frac{(\frac{\partial \sigma}{\partial S})^2 + (\frac{\partial \sigma}{\partial t})^2 + \lambda H_1(\sigma)}{\frac{\partial \sigma}{\partial t}}$

  The $\frac{\partial \sigma}{\partial S}^2$ and $H_1(\sigma)$ are constants with respect to $\frac{\partial \sigma}{\partial t}$, then we have:

$\frac{\partial}{\partial t}\frac{\partial L}{\partial \sigma'_t} = \frac{\partial}{\partial t}(\frac{\partial \sigma}{\partial t})^2 \Rightarrow \boxed{\frac{\partial}{\partial S}\frac{\partial L}{\partial \sigma'_S} = 2\frac{\partial^2 \sigma}{\partial t^2}.}$

- The last term is $\frac{\partial L}{\partial \sigma}$, we have: $\frac{\partial L}{\partial \sigma} = \frac{\partial((\frac{\partial \sigma}{\partial S})^2 + (\frac{\partial \sigma}{\partial t})^2 + \lambda H_1(\sigma))}{\partial \sigma}$

  $\frac{\partial L}{\partial \sigma} = \frac{\partial(\lambda H_1(\sigma))}{\partial \sigma}$

  Then the term $\frac{\partial L}{\partial \sigma}$ is given as:

$$\boxed{\frac{\partial L}{\partial \sigma} = 2\lambda \sum_i \sum_j (U_{ij}(\sigma) - \widetilde{U}_{ij})(\frac{\partial U}{\partial \sigma})}$$

Then, the new equation to solve is:

$$\frac{\partial^2 \sigma}{\partial S^2} + \frac{\partial^2 \sigma}{\partial t^2} = \lambda W_1(\sigma). \tag{3.38}$$

Where:

$$W_1(\sigma) = \sum_{i=1}^{N_1}\sum_{j=1}^{M_1} \frac{\partial U}{\partial \sigma}(y_j, T_i, \sigma)(U(y_j, T_i, \sigma) - U_{ij}) \tag{3.39}$$

computation of $\frac{\partial U}{\partial \sigma}$:,

We consider the partial differential operator :

$$\mathfrak{L}(\sigma) = \frac{\partial}{\partial T} + (r - q + \frac{1}{2}\sigma^2(y,T))\frac{\partial}{\partial y} - \frac{1}{2}\sigma^2(y,T)\frac{\partial^2}{\partial y^2} + q \tag{3.40}$$

Then if we take any suitably smooth function h, we may formally differentiate the expression:

$$\mathfrak{L}(\sigma + \varepsilon h)U(y, T, \sigma + \varepsilon h) = 0$$

with respect to ε. We Obtain

$$h\frac{\partial}{\partial \sigma}\frac{\partial U}{\partial T} + h(r - q + \frac{1}{2}\sigma^2(y,T))\frac{\partial}{\partial \sigma}\frac{\partial U}{\partial y} + h\sigma(y,T)\frac{\partial U}{\partial y} - \frac{h}{2}\sigma^2(y,T)\frac{\partial}{\partial \sigma}\frac{\partial^2 U}{\partial y^2} - h\sigma(y,T)\frac{\partial^2 U}{\partial y^2} + hq\frac{\partial U}{\partial \sigma} = 0$$

Equivalent to:

$$\frac{\partial}{\partial \sigma}\frac{\partial U}{\partial T} + (r - q + \frac{1}{2}\sigma^2(y,T))\frac{\partial}{\partial \sigma}\frac{\partial U}{\partial y} + \sigma(y,T)\frac{\partial U}{\partial y} - \frac{1}{2}\sigma^2(y,T)\frac{\partial}{\partial \sigma}\frac{\partial^2 U}{\partial y^2} - \sigma(y,T)\frac{\partial^2 U}{\partial y^2} + q\frac{\partial U}{\partial \sigma} = 0$$

We reverse the order of differentiation this leads to the equation

$$(\frac{\partial}{\partial T}+(r-q+\frac{1}{2}\sigma^2(y,T))\frac{\partial}{\partial y}-\frac{1}{2}\sigma^2(y,T)\frac{\partial^2}{\partial y^2}+q)\frac{\partial U}{\partial \sigma}=\sigma(y,T)(\frac{\partial^2 U}{\partial y^2}-\frac{\partial U}{\partial y})$$

Finally, $\frac{\partial U}{\partial \sigma}$ is writing as:

$$\mathfrak{L}(\sigma)\frac{\partial U}{\partial \sigma}=\sigma(y,T)(\frac{\partial^2 U}{\partial y^2}-\frac{\partial U}{\partial y}) \quad (3.41)$$

3.3.3 Algorithms and Numerical tests

In this section, we describe briefly the proposed Algorithms and the obtained results.

3.3.3.1 Algorithm 1

In Algorithm 1, we describe the steps to solve the elliptic PDE (3.38).
The calibration procedure require the numerical solution of PDE (3.38), and (3.7), (3.41) is the intermediate PDE. This can be accomplished conveniently using the finite-difference technique. In order to the numerical approximation using Finite difference technique we will consider a bounded domain $[y_{min}, y_{max}] \times [0, T_{max}]$. The numerical procedure to minimize (3.34) and determine σ can now be summarized by the following algorithm.

- **Numerical Results**

To test this Algorithm we choose the example used in [34]. We consider the true volatility function $\sigma_{true}(S) = 0.2 + 0.1e^{-0.2S}$, we use this function for generate a set of option prices U^{ij} by solving Dupire equation(3.7), the interest rate $r = 0.05$ we consider $T_{max} = 10$, and the strike $K \in [50, 160]$. We present the L^2 norm of the difference between successive iterations σ_k and σ_{k+1} and the maximum difference between the true value of σ and the final approximation over the given domain. In the table 3.1, we illustrate the different results of the proposed Algorithm and the calibration Algorithm [34] using the Black-Scholes.

Comparison: We note that the results obtained using the algorithm 1 are better than

Algorithm 1 Computation of σ using equation (3.38)

Require: $S \in [S_{min}, S_{max}]$, $t \in [0, T_{max}]$, Where T_{max} is the longest time expiration.
1. Select a finite-difference grid for Poisson equation $\{(S_m, t_n) : m = 0, .., M$ and $n = 0, .., N\}$
2. Choose an initial value of σ_0 (approximation of the true volatility, e.g $\sigma_0(S_m, t_n) = constant$)

for $m = 0$ to M **do**

  for $n = 0$ to N **do**

   3. Solving the Dupire PDE using the finite-difference method to compute the:$U(y, T, \sigma_0(S_m, t_n))$

$$\frac{\partial U}{\partial T}(y,T) = -qU(y,T) - (r - q + \frac{1}{2}\sigma^2(y,T))\frac{\partial U}{\partial y}(y,T) + \frac{1}{2}\sigma^2(y,T)\frac{\partial^2 U}{\partial y^2}(y,T)$$

   4. Using the finite-difference method we calculate $\frac{\partial U(y,t,\sigma)}{\partial \sigma}$ (equation((3.41))

5. Calculation of : $W_1(\sigma) = \sum_{i=1}^{N_1}\sum_{j=1}^{M_1}\frac{\partial U}{\partial \sigma}(y_j, T_i, \sigma)(U(y_j, T_i, \sigma) - U_{ij})$.
6. Solution of the Poisson equation:

$$\frac{\partial^2 \sigma}{\partial S^2} + \frac{\partial^2 \sigma}{\partial t^2} = \lambda W_1(\sigma),$$

7. Take the volatility function obtained in step 6, named σ_1,

if difference between σ_1 and σ_0 is negligible **then**

  return σ_0, **Stop**

else $\{\sigma_0 \leftarrow \sigma_1\}$

  return to Step 3.

return σ

the results found by the algorithm proposed by Chiarella et al. [34].

The Algorithm proposed in this paper permits to obtain an error of 10^{-4} in approximately 50 % of the time used by the algorithm proposed in [34] for only an error of 10^{-3}. The limitations of both algorithms, is that the numerical value of the regularization parameter λ differs from the range of theoretical values of this parameter. The convergence of both programs is obtained for a value λ that does not exceed one. We suggest the improvement of this algorithm, using a false time parameter as proposed by Lagnado [93] in order to use more important value of the parameter λ.

		Calibration volatility with BS	Algorithm 1
λ=0.1, σ_0=0.2	N^o.Iteration & Time.Comp (s)	20 & 312	8 & 175
	$\|\sigma_{k+1}-\sigma_k\|_2$	0.92432e-06	0.0284e-06
	$\max\|\sigma_{true}-\sigma_{k+1}$	0.00368	0.000192
λ=0.2, σ_0=0.1	N^o.Iteration & Time.Comp (s)	21 & 328	12 & 184
	$\|\sigma_{k+1}-\sigma_k\|_2$	0.84226e-06	0.09325e-06
	$\max\|\sigma_{true}-\sigma_{k+1}$	0.00422	0.00109
λ=1, σ_0=0.2	N^o.Iteration & Time.Comp (s)	25 & 344	16 & 202
	$\|\sigma_{k+1}-\sigma_k\|_2$	1.26932e-06	1.00633e-06
	$\max\|\sigma_{true}-\sigma_{k+1}$	0.00625	0.00342
λ=10, σ_0=0.1	N^o.Iteration & Time.Comp (s)	set on 20 & –	set on 20 & –
	$\|\sigma_{k+1}-\sigma_k\|_2$	–	–
	$\max\|\sigma_{true}-\sigma_{k+1}$	12.45039	14.99526

Table 3.1: Comparison results:calibration σ with B-S and algorithm 1

3.3.3.2 Algorithm 2

In this Algorithm, we introduce a 'false' time parameter θ and a function $\widehat{\sigma}(S,t,\theta)$. Starting with some initial guess $\widehat{\sigma}(S,t,\theta)$, we will solve a perturbed parabolic equation under the form :

$$\frac{\partial\widehat{\sigma}}{\partial\theta}=\frac{\partial^2\widehat{\sigma}}{\partial S^2}+\frac{\partial^2\widehat{\sigma}}{\partial t^2}-\lambda W_1(\sigma) \tag{3.42}$$

Where $W_1(\sigma)$ is given by:

$$W_1(\sigma)=\sum_{i=1}^{N_1}\sum_{j=1}^{M_1}\frac{\partial U}{\partial\sigma}(y_j,T_i,\sigma)(U(y_j,T_i,\sigma)-U_{ij}) \tag{3.43}$$

Therefore, instead of solving an elliptical equation, we consider a parabolic equation. The new algorithm to compute σ using the equation (3.42) is presented in the following.

- **Numerical Results**
  For a numerical tests we go back to the previous example. To examine the effectiveness of the Algorithm 2, in regard to the parameter λ, we give for λ a higher value. We use $\Delta\theta$ =0.0001 for the step discretization from the false time. The comparison of the three Algorithm is given in table 3.2.

Algorithm 2 Algorithm to compute σ using equation (3.42)

Require: $S \in [S_{min}, S_{max}]$, $t \in [0, T_{max}]$, Where T_{max} is the longest time expiration.
1.Select a finite-difference grid for Parabolic equation $\{(S_m, t_n) : m = 0, .., M$ and $n = 0, .., N\}$,
2. Choose an initial value of σ_0 (approximation the true volatility,e.g $\sigma_0(S_m, t_n) = constant$),
for $m = 0$ to M **do**
for $n = 0$ to N **do**
3. Solving the Dupire PDE using the finite-difference method to compute the: $U(y, T, \sigma_0(S_m, t_n))$

$$\frac{\partial U}{\partial T}(y,T) = -qU(y,T) - (r - q + \frac{1}{2}\sigma^2(y,T))\frac{\partial U}{\partial y}(y,T) + \frac{1}{2}\sigma^2(y,T)\frac{\partial^2 U}{\partial y^2}(y,T)$$

4. Using the finite-difference method we calculate $\frac{\partial U(y,t,\sigma)}{\partial \sigma}$ (equation(3.41))
5. Calculation of : $W_1(\widehat{\sigma}) = \sum_{i=1}^{N_1}\sum_{j=1}^{M_1} \frac{\partial U}{\partial \sigma}(y_j, T_i, \sigma)(U(y_j, T_i, \sigma) - U_{ij})$
6. Solution of parabolic equation:

$$\frac{\partial \widehat{\sigma}}{\partial \theta} = \frac{\partial^2 \widehat{\sigma}}{\partial S^2} + \frac{\partial^2 \widehat{\sigma}}{\partial t^2} - \lambda W_1(\sigma)$$

For a small increment $\Delta\theta$ to find updated values of $\widehat{\sigma}$ at all interior points.
7.
if difference between $\widehat{\sigma}_1$ and $\widehat{\sigma}_0$ is smaller than some desired tolerance. **then**
return $\widehat{\sigma}_0$, **Stop**
else $\{\widehat{\sigma}_0 \leftarrow \widehat{\sigma}_1\}$
return to Step 3.
return $\widehat{\sigma}$

		Algorithm with BS	Algorithm 1	Algorithm 2
σ_0=0.2	N^o.Iteration & Time.Comp (s)	25 & 344	16 & 202	20 & 239
λ=1	$\mid \sigma_{k+1} - \sigma_k \mid_2$	1.26932e-06	1.00633e-06	3.69032e-06
	$\max \mid \sigma_{true} - \sigma_k \mid$	0.00625	0.00342	0.045228
σ_0=0.1	N^o.Iteration & Time.Comp (s)	20 & –	20 & –	20 & 201
λ=10	$\|\sigma_{k+1} - \sigma_k\|_2$	–	–	1.02988e-06
	$\max \mid \sigma_{true} - \sigma_k \mid$	12.45039	14.99526	0.002892

Table 3.2: Comparison of three algorithm

Comparison: We note that the Algorithm 2, converges for the higher value of λ and

	N^o.Iteration & Time.Comp (s)	$\|\sigma_{k+1} - \sigma_k\|_2$	$\max \|\sigma_{true} - \sigma_k\|$
σ_0=0.1, and λ=30	17 & 170	0.18632e-06	0.000078
σ_0=0.2, and λ=30	16 & 165	0.09918e-06	0.000021
σ_0=0.1, and λ=100	14 & 152	0.02877e-06	0.000014
σ_0=0.2, and λ=100	13 & 149	0.02243e-06	0.000012

Table 3.3: Results of Algorithm 2

gives a better results. So, it satisfies the theoretical point of view of the regularization parameter, and gives an error that is less than 10^{-4}. In the next section we suggest to improve this algorithm at the level of the initial guess of the volatility, using a statistical approximation.

3.3.3.3 Algorithm 3

In the Algorithm 3, we are interested in the Computation of *sigma* using statistical approximation of the initial volatility.
We will use the statistical approximation used in [79]. First, we denote:

- $n+1$: observations number.
- S_i : the underlying in the time i.
- τ : duration time intervals

The estimation s, is given by the standard deviation:

$$s = \sqrt{\frac{1}{n}\sum_{i=1}^{n}(u_i - \overline{u})^2} \tag{3.44}$$

where : $u_i = ln(S_i/S_{i-1})$, for $i = 2, .., n$; and $\overline{u}$ is the average of u_i
Therefore the estimation of σ is:

$$\boxed{\sigma^* = s \setminus \sqrt{\tau}} \tag{3.45}$$

The new algorithm is written as follow :

- **Numerical results**
  - **First Test:**

Algorithm 3 Compute σ using statistical approximation of the initial volatility

Require: $S \in [S_{min}, S_{max}]$, $t \in [0, T_{max}]$, Where T_{max} is the longest time expiration.
1. Calculate the statistical σ^* using (3.44) and (3.45)
2. Select a finite-difference grid for Parabolic equation $\{(S_m, t_n) : m = 0, .., M$ and $n = 0, .., N\}$

for $m = 0$ to M **do**
 for $n = 0$ to N **do**
  3. Solving the Dupire PDE using the finite-difference method to compute the: $U(y, T, \sigma_0(S_m, t_n))$

$$\frac{\partial U}{\partial T}(y,T) = -qU(y,T) - (r - q + \frac{1}{2}\sigma^2(y,T))\frac{\partial U}{\partial y}(y,T) + \frac{1}{2}\sigma^2(y,T)\frac{\partial^2 U}{\partial y^2}(y,T)$$

  4. Using the finite-difference method to calculate $\frac{\partial U(y,t,\sigma)}{\partial \sigma}$(equation(3.41))
5. Calculation of : $W_1(\widehat{\sigma}) = \sum_{i=1}^{N_1} \sum_{j=1}^{M_1} \frac{\partial U}{\partial \sigma}(y_j, T_i, \sigma)(U(y_j, T_i, \sigma) - U_{ij})$.
6. Solving a discrete form of (3.42) for a small increment $\Delta\theta$ to find updated values of $\widehat{\sigma}$ at all interior points.

if difference between $\widehat{\sigma}_1$ and $\widehat{\sigma}_0$ is negligible. **then**
 return $\widehat{\sigma}_0$, **Stop**
else $\{\widehat{\sigma}_0 \leftarrow \widehat{\sigma}_1,\}$
 return to Step 3.
return $\widehat{\sigma}$

We take the previous example treated in subsection 3.3.3.1. To compute the volatility, we use algorithm 3. In the first, using the Statistical estimation of σ_0^*, we obtain $\sigma_0^* = 0.205$, after that, we take this value of σ_0^* in order to compute σ using Algorithm 3. We obtain after 15 iterations with $\lambda = 30$ an error max $| \sigma_{true} - \sigma_k | = 0.215\ 10^{-4}$.

– **Second Test:**

In this test, we use the example used in [25] (see page 20), the true volatility is:

$$\sigma_{true}(S,t) = \frac{15}{S}$$

These data corresponds to European call with: $S_0 = 100$; $r = 0.05$; $q = 0.02$.

In the numerical solution of the Dupire PDE, we choose the following parameters:

1. $[K_{min}; K_{max}] = [0.005; 200]$(then $[y_{min}; y_{max}]$=$[-5.298; 5298]$)

2. $[T_0; T_{max}] = [0; 1]$

We take $\lambda = 30$, after 16 iterations , and using the third Algorithm, the obtained results are as follows: For the Statistical estimation of σ_0^* we obtain, $\sigma_0^* = 0.1478$ The true volatility and the estimated one using Algorithm 3 are given in the figure 3.1.

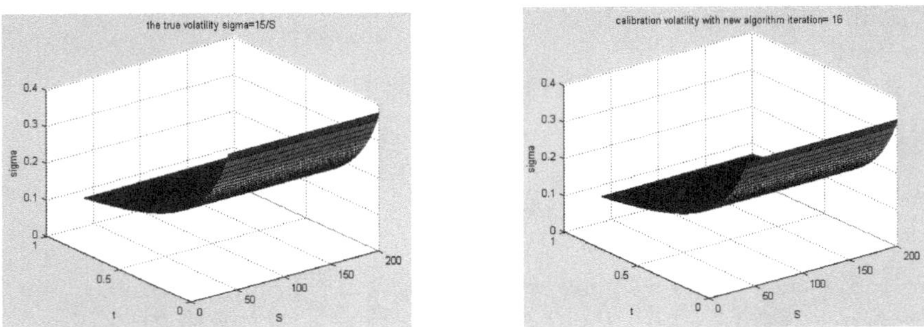

Figure 3.1: local volatility

– **Third Test:**

The third application can be demonstrated by carrying out the calibration using index option prices observed in the market. We selected the closing prices of some $S\&P500$ index options as of 3 January 2011 for this test. We choose the options expiring in 25 May 2011. We take $\lambda = 30$, the results obtained after 16 iterations using Algorithm 3 was compared with the true volatility. The statistical approximation of the initial guess volatility is $\sigma_0^* = 22.45\%$.

3.4 Comparison Alternative Regularization with Jump indicator Model

This section is devoted to a comparison of the Alternative Regularization Method presented in the previous chapter and the Jump Indicators method proposed in the work of Aboulaich et al. [6].

Firstly, we briefly present the jump indicators method for estimation of the volatility function, assuming it is piecewise constant of time. This method was initially developed for the estimation of hydraulic transmissivity in porous media in [21]. Then, the Algorithm for volatility estimation using

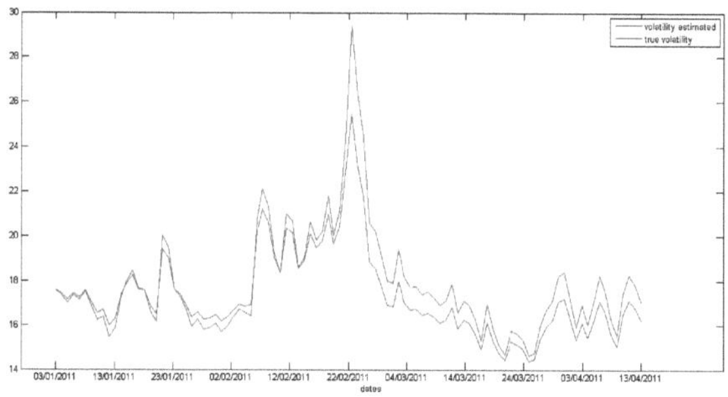

Figure 3.2: comparison of the observed volatility in the market and that estimated by algo3

jump indicators is given.

Finally, we introduce the numerical results of our Algorithm presented in 3 and the Algorithm using jump indicators method, to compare the effectiveness of both algorithms in terms of accuracy.

3.4.1 Jump Indicator for estimation volatility

The volatility estimate is an inverse problem can be formulated as a minimization problem of an objective function defined as the difference between the market data and the corresponding quantities calculated assuming the current value of the parameter p. We define a jump indicator in order to define a partition of the domain :

$$\Omega = \{(S, M, t) \in [0, 1] \times [0, 1] \times [0, T]/S \leq M\}$$

given by:

$\Omega_1 = \{(S, M, t) \in [0, 1]\times[0, 1]\times[0, t_1]/S \leq M\}$ and $\Omega_2 = \{(S, M, t) \in [0, 1]\times[0, 1]\times[t_1, T]/S \leq M\}$ Ω_k, $k = 1, 2$.
such that the volatility is constant in each part . The method is based on a computation of an indicator that permits to locate iteratively the jumps. This indicator locates the volatility discontinuities which has the greater amplitude and more effect to the underlying and step by step they will try to locate the smaller amplitude jumps.

Let (P_1) be the minimization problem considering the volatility is constant during the maturity. Therefore, a single parameter σ_0^* will be estimated and that corresponds to

the optimal solution of the minimization problem $J(\sigma)$. Let be (P_2), the minimization problem of $J(\sigma)$ associated with a new partition z_1 and z_2 supposing that σ has one discontinuity. The solution of the problem (P_2) is presented as $J_{P_2}(\sigma_1^*, \sigma_2^*)$ with σ_1^* corresponds to the area z_1 and σ_2^* corresponds to the area z_2, we note $\sigma^* = (\sigma_1^*, \sigma_2^*)$.
To determine the discontinuity, we will consider the difference c between the two parameters σ_1 and σ_2 and we consider the problem of minimization J_{P_2} under constraint $c = \sigma_2^* - \sigma_1^*$.

$$J_{P_2}(\sigma^*) = min_c J_{P_2}(\sigma) \tag{3.46}$$

If $c = 0$, then the solution $J_{P_2}(\sigma^*)$ (with $\sigma^* = (\sigma_0^*, \sigma_0^*)$) corresponds to the first problem. In the other case, the optimal value of the problem under constraint c corresponds to the optimum value of $J_{P_2}(\sigma_1^*, \sigma_2^*)$.
if we note $J^c = J_{P_2}(\sigma_1^*, \sigma_2^*)$ and $J^0 = J_{P_2}(\sigma_0^*, \sigma_0^*)$, using:

$$J^c = J^0 + \frac{\partial J^c}{\partial c} \mid_{c=0} .c + O(c) \tag{3.47}$$

The norm $\| \frac{\partial J^c}{\partial c} \mid_{c=0} \|$ measures the variation in J^c in relation to the difference between the two parameters σ_1^* and σ_2^* through c, this norm is the Jump Indicator.
On the other hand, we associate the minimization problem under constraint $c = \sigma_2^* - \sigma_1^*$ to the Lagrangian defined by:

$$L_c(\sigma, \lambda) = J(\sigma) + < \lambda, A\sigma - c >$$

With $A = [-1.1]$ and λ is the Lagrange multiplier associated with the constraint $c = \sigma_2^* - \sigma_1^*$.
The optimum value $\sigma^* = (\sigma_1^*, \sigma_2^*)$ verifies the following conditions.
At the optimum, we have:

$$J_c^* = J(\sigma^*) = L_c(\sigma^*, \lambda^*), \quad and \quad \frac{\partial L_c}{\partial c} \mid_{c=0} (\sigma^*, \lambda^*) = 0 \tag{3.48}$$

Then:

$$\frac{\partial J_c^*}{\partial c} \mid_{c=0} - \lambda^* = 0 \tag{3.49}$$

3.4.2 Algorithm and Numerical Results

3.4.2.1 Algorithm for the jump determination

In this part, we present the Algorithm given in the work of Aboulaich et al. in [6]. Considering:
$\Omega = \Omega_1 \times [0, T]$ with $\Omega_1 = \{(x, y) \in [0, 1] * [0, 1]/x \leq y\}$ in two sub-domains $\Omega_1 \times [0, t_1]$

and $\Omega_1 \times [t_1, T]$, with t_1 is the moment where the norm $\| \frac{\partial J^c}{\partial c} \|$ is maximum.
We have for each time t_k, the Jump Indicator I_k is given by:

$$I_k = |(\sum_{t_l=0}^{k} \sum_{j=0}^{n} (\sum_{i=j}^{n} (\frac{\partial J}{\partial \sigma})_{x_i, y_j, t_l}))|$$

where

- $x_i = ih$; $y_j = jh$
- $h = \Delta x = \Delta y$
- and $t_l = l\Delta\text{t}$

The algorithm is then given by:

Algorithm 4 Algorithm 1: Estimation Volatility for one jump

- **For** $k = 1$ to a maximum time step.
- Resolution of the state equation to determine P;
- Calculation of ∇J^* from the condition of optimality.
- Computation of I_k.
- Return the Jump Indicator that corresponds to the max of I_k, and calculate σ_1

The generalization for the multiple jump the Algorithm is rewritten as follows:

3.4.2.2 Numerical Results

We generate prices from Black-Scholes model using fictive data:

$$S_t = S_0 \exp([r - \frac{\sigma^2}{2}]t + \sigma \varepsilon \sqrt{dt})$$

where ε follows the normal process.
The true volatility function is given by:

$$\sigma(S,t) = \begin{cases} 0.8 & if \quad 0 \leq t \leq 36 \\ 0.2 & if \quad 37 \leq t \leq 48 \\ 1.1 & if \quad 49 \leq t \leq 100 \end{cases}$$

Algorithm 5 Algorithm 2: Estimation Volatility for multiple jump

- **Repeat**
  - **For** $h = 1$ to n
  - Call Algorithm 4 and obtained $I_{k,h}$ and $\sigma_{k,h}$;
  - Calculate $J_h(\sigma_{k,h})$;
  - return σ_{k,h^*} and I_{k,h^*}) who gives the minimum of J_h;
- $I_{k+1} = I_{k,h^*}$ and $\sigma_{k+1} = \sigma_{k,h^*}$;
- while $\|\sigma_{k+1} - \sigma_k\| < \varepsilon$ or $\|I_k\| < \varepsilon$

We apply the Algorithm 3 of the Alterative Regularization method introduced in chapter 3, for $\lambda = 30$, the statistical approximation of the initial volatility is equal to: $\sigma_0^* = 0.572$. After 14 iteration we obtain the result. For the Algorithm 5 of the jump indicator method introduced in this chapter, we consider partition into three zones Z_1, Z_2, Z_3 with σ constant on each subset Z_i. We applied the method above to find these two discontinuities and calculate σ.
The results of comparison for the both method as shown in the table and figure bellow:

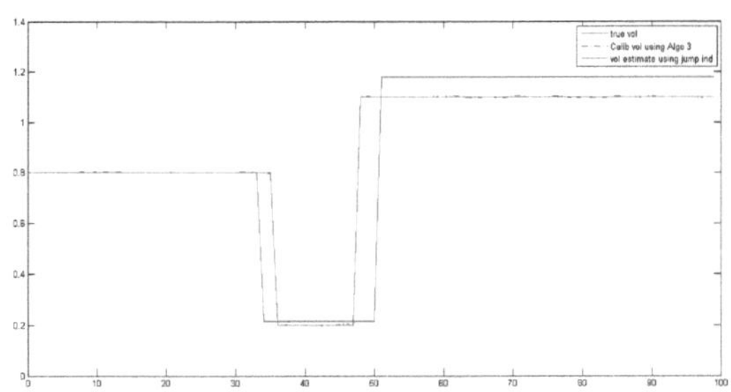

Figure 3.3: comparison of the volatility estimated by Alternative Regularization and indicator jump

Comparison The volatility estimate using Algorithm 3 gives a first jump on the same day of the jump, while the second jump delayed one day. The error of the estimate of volatility is 0.48% , and the CPU time recorded is 142 seconds.

	Volatility given	Volatility estimated by Regularization Method	Volatility estimated by Jump Indicator
σ_1	0.8	0.8000	0.8
σ_2	0.2	0.2003	0.216
σ_3	1.1	1.0998	1.18

Table 3.4: Comparison of the volatility obtained by the regularization method and the indicator jump

The calculation of indicator gives us a first jump on the day 35 two days as error before the exact date while estimated volatility error is around 8%. Detection of the second jump is delayed by 3 days compared to the exact date when the error of the estimate of the volatility of 7 CPU time recorded is 29 seconds. If the jumps indicator method is faster than the time of compute σ using the Algorithm 3, we note that the algorithm 3 gives more accurate estimation of the value of the volatility.

3.5 Conclusion

The estimated volatility function was the aim of the first part of the book. We tried to provide an algorithm that allows to have a compromise between accuracy and computation time.

We proposed an algorithm which is inspired by Lagnado in [93] and Chiarella et al. in [34] methods, but differs from these methods in the resolution of the direct problem for pricing Option.
In our method the price is calculated from the EDP Dupire with logarithmic change instead of Black Scholes PDE, which optimizes with respect to computation time. Using a Euler- Lagrange equation, the computation of the volatility is reduced to a resolution of a Parabolic equation as proposed in [93].

In order to improve the method we propose to use a statistical estimation for the computation of the initial guess of a volatility.
The numerical experiments presented in section 3.3.2 and 3.4.1 prove the efficiency of the proposed Algorithm.

Chapter 4

Pricing Option with Runge-Kutta Direct Discontinuous Galerkin Method

---

Collaborated with : **R. Aboulaich**[1], **O. Bokanowski**[2]

Contents

The discontinuous Galerkin method is a class of finite element methods using completely discontinuous piecewise polynomial space for the numerical solution and the test functions. The numerical fluxes is the key ingredient of this method to obtain highly accurate and stable schemes in many difficult

[1]EMI,Mohammed V University, Rabat

[2]Paris 7 University, Paris

situations.

The first discontinuous Galerkin (DG) method originates from Reed and Hill [109] in 1973, who apply this method to the neutron transport equation, i.e, a time independent linear hyperbolic equation. A error estimation of the DG method is given by Lesaint & Raviart (1974) [94]. A major development of the DG method was performed by Cockburn, Shu et al. in a series of papers [45, 46, 44, 47, 49], in which they have established a framework to easily solve nonlinear time-dependent problems with explicit Runge-Kutta methods for the time- discretization and DG discretization in Space with exact or approximate Rieman solvers as interface fluxes and Total Variation Bounded (TVB). These schemes are termed RKDG methods. They also introduce the concepts of local projection limiters for resolving sharp gradients, such as shocks, without oscillations, and characteristic decomposition and splitting of element boundary fluxes. Recently, Zhang & Shu [129] have constructed uniformly high-order accurate schemes satisfying a strict maximum principal for scalar conservation laws. A pedagogical presentation of the DG discretization can be found in [43, 116, 42]. The extension of the local discontinuous Galerkin methods to partial differential equation in [32].

The highlight of the DG methods that combine noteworthy advantage from the Finite Element methods and the Finite Volume. First, the high accuracy comes from the high-order polynomial approximation of the solution. This approximation results from a local variational formulation on each cell. Then, this polynomial description of the solution being discontinuous from one cell to another, the fluxes at the interfaces are evaluated by means of numerical flux functions similar to the ones used in the Finite Volume method frame.

The DG method has found rapid application in such diverse areas as aeroacoustic, electro-magnetism, magnetohydrodynamics, turbulent flows, viscoelastic, among many others. We are interested in this chapter for applications in finance, precisely pricing European and American Option. We start with discussion and analysis the RKDDG method for one dimensional scalar conservation laws. In the second section, we developed RKDDG method for Diffusion-advection problem and we present the scheme RKDDG for European Option Equation. Then, the last section is devoted for pricing American Option.

4.1 One-dimensional scalar conservation laws

This section focuses to the description of numerical methods to solve the following problem:

$$\begin{cases} \partial_t u(x,t) + a(x)\partial_x u(x,t) = 0 & \forall (t,x) \in [0,T] \times [x_{min}, x_{max}], \\ u(x,0) = u_0(x) & \forall x \in [x_{min}, x_{max}]. \end{cases} \tag{4.1}$$

Let $u = u(x,t)$, for $x \in [x_{min}, x_{max}] \subset \mathbb{R}$, and $t \in [0,T]$, be the solution of One-Dimensional scalar conservation law, and assume, for being, that a(x) is a smooth function.

4.1.1 Space discretization

We shall first consider a semi-discretization with respect to the space variable.
For this, we consider a subdivision of the interval $[x_{min}, x_{max}]$ on I cells
$C_i = [x_{i-\frac{1}{2}}, x_{i+\frac{1}{2}}]$, for $i = 1, ..., I$, where,

$$x_{min} = x_{\frac{1}{2}} < x_{\frac{3}{2}} < \cdots < x_{I+\frac{1}{2}} = x_{max}$$

We again denote:

$$x_i = \frac{1}{2}(x_{i-\frac{1}{2}} + x_{i+\frac{1}{2}}), \qquad \Delta x_i = x_{i+\frac{1}{2}} - x_{i-\frac{1}{2}}, \qquad \Delta x = \max_i \Delta x_i, \qquad \tau = \Delta t$$

where, h is the Space step, and τ is the time step.
The cells do not need to be uniform for the method, but for simplicity of analysis we will take the uniform mesh, $\Delta x_i = cst$.
Let $K = (K_1, K_2, \cdots, K_I)$ is the vector where each component K_i represents the highest degree of the polynomial in the cell C_i. For each integer $K_i \geq 1$, we define the approximation space as

$$V_h^{K_i} = \{v : v|_{C_i} \in \mathbf{P}^{K_i}(C_i), i = 1, ..., I\}. \tag{4.2}$$

where $\mathbf{P}^{K_i}(C_i)$ is the space polynomials in C_i of degree at most K_i. In this work, we suppose all polynomials of degree at most K.($K = K_i$, for each $i \in 1, ..., I$). If u_h denotes the polynomial approximation of the solution, let u_h^i be the restriction of u_h, over the cell C_i

$$u_h^i(x,t) \simeq \sum_{k=0}^{K} u_k^i(t)\varphi_k^i(x), \qquad x \in C_i \tag{4.3}$$

where, u_k^i are the successive moments of u_h and $\{\varphi_k^i\}_{k=0,...,K}$ is basis of $\mathbf{P}^K(C_i)$, with $\varphi_k^i(x) = (\frac{x-x_i}{h})^k$

4.1.2 Formulation of the scheme

First, we multiply the PDE (4.1) by a "test function" v, then integrate the result over the interval C_i,

$$\int_{C_i}(\frac{\partial u}{\partial t}(x,t)+a(x)u(x,t)\partial_x)v(x)dx=0.$$

Now consider u and v both from a finite-dimensional function space V_h^k. Then perform integration by parts and write

$$\int_{C_i}\frac{\partial u}{\partial t}(x,t)v(x)dx-\int_{C_i}a(x)\partial_x u(x,t)v_h(x)dx+\underbrace{\hat{u}_{i+\frac{1}{2}}a(x_{i+\frac{1}{2}})v(x_{i+\frac{1}{2}})-\hat{u}_{i-\frac{1}{2}}a(x_{i+\frac{1}{2}})v(x_{i-\frac{1}{2}})}_{?}$$

$$=\int_{C_i}\partial_x a(x)u(x,t)v(x)dx \qquad (4.4)$$

is true for any test function $v\in V_h^k$, but the term marked "?" is meaningless, since the function are double-valued at the spots in question. To motivate a meaning for the term, consider the following: if we take the test function

$$v=\begin{cases}1 & x\in \quad C_i\\ 0 & elsewhere,\end{cases}$$

we recover

$$\int_{C_i}\frac{\partial u}{\partial t}(x,t)dx-\widehat{u}_{i+\frac{1}{2}}a(x_{i+\frac{1}{2}})\underbrace{v(x_{i+\frac{1}{2}})}_{from\quad left}-\widehat{u}_{i-\frac{1}{2}}a(x_{i+\frac{1}{2}})\underbrace{v(x_{i-\frac{1}{2}})}_{from\quad right}=\int_{C_i}\partial_x a(x)u(x,t)dx$$

Let us point that the function u_h may be discontinuous at the points $x_{i+\frac{1}{2}}$, then we replace the nonlinear 'flux' $u(t,x_{i+\frac{1}{2}})$ by a numerical 'flux' $\widehat{u}_{i+\frac{1}{2}}(t)$ is again the numerical flux, which is exactly reminiscent of the FV scheme, motivating the equality

$$\widehat{u}_{i+\frac{1}{2}}=(u(x^+_{i+\frac{1}{2}},t),u(x^-_{i+\frac{1}{2}},t))$$

and thus the scheme

$$\int_{C_i}\frac{\partial u}{\partial t}(x,t)v(x)dx-\int_{C_i}a(x)u(x,t)\partial_x v_h(x)dx+\widehat{u}_{i+\frac{1}{2}}a(x_{i+\frac{1}{2}})v(x^-_{i+\frac{1}{2}})$$

$$-\widehat{u}_{i-\frac{1}{2}}a(x_{i+\frac{1}{2}})v(x^+_{i-\frac{1}{2}})=\int_{C_i}\partial_x a(x)u(x,t)v(x)dx \qquad (4.5)$$

The best-know examples of numerical fluxes satisfying the above properties are the following:(We denote $f(u(x,t))=a(x)u(x,t)$)

1. The Lax-Friedrichs flux:

$$\widehat{f}^{LF}(u^-,u^+)=\frac{1}{2}(f(u^-)+f(u^+)-\alpha(u^+-u^-)),\ \alpha=\max_u\mid f'(u)\mid$$

2. The Godunov flux:

$$\widehat{f^G}(u^-,u^+) = \begin{cases} \min_{u^- \leq u \leq u^+} f(u), & if \quad u^- < u^+; \\ \max_{u^+ \leq u \leq u^-} f(u), & if \quad u^- \geq u^+; \end{cases}$$

3. The Engquist- Osher flux

$$\widehat{f^G}(u^-,u^+) = \int_0^{u^-} \max(f'(u),0)du + \int_0^{u^+} min(f'(u),0)du + f(0).$$

Following [48], we can define the upwind version of our scheme as: find $u_h \in V_h$, such that

$$\int_{C_i} (\frac{\partial u_h}{\partial t}(x,t) + a(x)\frac{\partial u_h}{\partial x}(x,t))v_h(x)dx + \frac{1}{2}(a(x_{i+\frac{1}{2}}) - \mid a(x_{i+\frac{1}{2}}) \mid)[u_h]_{i+\frac{1}{2}}(v_h)^-_{i+\frac{1}{2}}$$
$$+\frac{1}{2}(a(x_{i-\frac{1}{2}}) + \mid a(x_{i-\frac{1}{2}}) \mid)[u_h]_{i-\frac{1}{2}}(v_h)^+_{i-\frac{1}{2}} = 0, \tag{4.6}$$

where : $[u]_{i-\frac{1}{2}} = u^+_{i-\frac{1}{2}} - u^-_{i-\frac{1}{2}}$.

4.1.2.1 Discontinuous Galerkin Scheme for a(x)=cst

We start with the simplest case where the function $a(x)$ is constant, consider the Transport Equation:

$$\begin{cases} \partial_t u(x,t) + a\partial_x u(x,t) = 0 & (t,x) \in [0,T] \times [x_{min}, x_{max}], \\ u(x,0) = u_0(x) & \forall x \in [x_{min}, x_{max}]. \end{cases} \tag{4.7}$$

The formulation (4.7) is given by:

$$\begin{cases} \int_{C_i} (\partial_t u_h(x,t) + a\partial_x u_h(x,t))v_h(x) + \frac{1}{2}(a - \mid a \mid)[u_h]_{i+\frac{1}{2}}(v_h)^-_{i+\frac{1}{2}} \\ \qquad\qquad + \frac{1}{2}(a + \mid a \mid)[u_h]_{i-\frac{1}{2}}(v_h)^+_{i-\frac{1}{2}} = 0 \\ \int_{C_i} u(x,0)v_h(x)dx = \int_{C_i} u_0(x)v_h(x)dx. \end{cases} \tag{4.8}$$

If $a > 0$, the equation (4.8) is given by:

$$\int_{C_i} \partial_t u_h(x,t)v_h(x)dx + a\int_{C_i} \partial_x u_h(x,t)v_h(x)dx + a[u]_{i-\frac{1}{2}}(v_h)^+_{i-\frac{1}{2}} = 0 \tag{4.9}$$

where : $[u]_{i-\frac{1}{2}} = u^+_{i-\frac{1}{2}} - u^-_{i-\frac{1}{2}}$.
And, if $a < 0$ then:

$$\int_{C_i} \partial_t u_h(x,t)v_h(x)dx + \int_{C_i} a\partial_x u_h(x,t)v_h(x)dx + a[u]_{i+\frac{1}{2}}(v_h)^-_{i+\frac{1}{2}} = 0 \tag{4.10}$$

where : $[u]_{i+\frac{1}{2}} = u^+_{i+\frac{1}{2}} - u^-_{i+\frac{1}{2}}$.

Let $\{1, x, \cdots, x^K\}$, the basis of V_h, and let us approximate solution u_h:

$$u^i_h(x,t) \simeq \sum_{k=0}^{K} u^i_k(t)\varphi^i_k(x), \tag{4.11}$$

where, $\varphi^i_k(x) = (\frac{x-x_i}{h})^k$

Discrete Galerkin Scheme of order (K+1) in equation (4.6) becomes:
find $u_h \in V^K_h$ such as:

$$\int_{C_i} \partial_t u_h(t,x)v_h(x)dx + \int_{C_i} a\partial_x u_h(t,x)v_h(x)dx + a[u]_{i-\frac{1}{2}}(v_h)^+_{i-\frac{1}{2}} = 0 \tag{4.12}$$

Now take $v = \varphi^i_q(x)$, $q = 0, \ldots, K$ and put that into our scheme, then the equation (4.12) becomes as follows : $\forall i = 1, ..., I$ and $k = 0, ..., K$:

$$\sum_{k=0}^{K} \partial_t u^i_k(t) \int_{C_i} \varphi^i_k(x)\varphi^i_q(x)dx + a\sum_{k=0}^{K} u^i_k(t) \int_{C_i} \partial_x(\varphi^i_k(x))\varphi^i_q(x)dx + a[u]_{i-\frac{1}{2}}(\varphi^i_q)^+_{i-\frac{1}{2}} = 0 \tag{4.13}$$

Which leads to :

$$\underbrace{\sum_{k=0}^{K} M_{qk}\partial_t u_k(t)}_{(\mathbf{M}\frac{\partial \mathbf{u^i}}{\partial \mathbf{t}})_\mathbf{q}} + a\underbrace{\sum_{k=0}^{K} S_{qk}u_k(t)}_{(\mathbf{Su^i})_\mathbf{q}} + a[u]_{i-\frac{1}{2}}(\varphi^i_q)^+_{i-\frac{1}{2}} = 0 \tag{4.14}$$

where M and S are matrices of $\mathbb{R}^{(K+1)\times(K+1)}$ s.t:

$$M_{qk} = \int_{x_{i-\frac{1}{2}}}^{x_{i+\frac{1}{2}}} \varphi^i_k(x)\varphi^i_q(x)dx = \frac{\Delta x}{k+q+1}.\frac{1}{2^{k+q+1}}(1-(-1)^{k+q+1}). \tag{4.15}$$

$$S_{qk} = \int_{x_{i-\frac{1}{2}}}^{x_{i+\frac{1}{2}}} \partial_x\varphi^i_k(x)\varphi^i_q(x)dx = \frac{k}{k+q}.\frac{1}{2^{k+q}}(1-(-1)^{k+q}). \tag{4.16}$$

If $\mathbf{a > 0}$, the last term $[u]_{i-\frac{1}{2}}v^+_{i-\frac{1}{2}}$ in (4.9) , can be computed as follows:

$$[u]_{i-\frac{1}{2}}v^+_{i-\frac{1}{2}} = u^+_{i-\frac{1}{2}}(t)v^+_{i-\frac{1}{2}} - u^-_{i-\frac{1}{2}}(t)v^+_{i-\frac{1}{2}} \tag{4.17}$$

We take $v = \varphi^i_q$ in formula (4.13), and we have :

$$u^+_{i-\frac{1}{2}}(t)v^+_{i-\frac{1}{2}} - u^-_{i-\frac{1}{2}}(t)v^+_{i-\frac{1}{2}} = \sum_{k=0}^{K} u^i_k(t)\underbrace{\varphi^i_k(x^+_{i-\frac{1}{2}})\varphi^i_q(x^+_{i-\frac{1}{2}})}_{P^i_{qk}} - \sum_{k=0}^{K} u^{i-1}_k(t)\underbrace{\varphi^{i-1}_k(x^-_{i-\frac{1}{2}})\varphi^i_q(x^+_{i-\frac{1}{2}})}_{Q^i_{qk}}$$

$$= (Pu^i)_q - (Qu^{i-1})_q$$

where :

$$\mathbf{P^i_{qk}} = (-\frac{\mathbf{1}}{\mathbf{2}})^{\mathbf{k+q}} \quad and \quad \mathbf{Q^i_{qk}} = \frac{(-\mathbf{1})^{\mathbf{q}}}{\mathbf{2^{k+q}}} \tag{4.18}$$

If a < 0, the last term $[u]_{i+\frac{1}{2}}v^-_{i+\frac{1}{2}}$ in (4.10), we get:

$$[u]_{i+\frac{1}{2}}v^-_{i+\frac{1}{2}} = u^+_{i+\frac{1}{2}}(t)v^-_{i+\frac{1}{2}} - u^-_{i+\frac{1}{2}}(t)v^-_{i+\frac{1}{2}} \tag{4.19}$$

Then the formula (4.19) is written as:

$$u^+_{i+\frac{1}{2}}(t)v^-_{i+\frac{1}{2}} - u^-_{i+\frac{1}{2}}(t)v^-_{i+\frac{1}{2}} = \sum_{k=0}^{K} u^{i+1}_k(t)\underbrace{\varphi^i_k(x^+_{i+\frac{1}{2}})\varphi^i_q(x^-_{i+\frac{1}{2}})}_{E^i_{qk}} - \sum_{k=0}^{K} u^i_k(t)\underbrace{\varphi^i_k(x^-_{i+\frac{1}{2}})\varphi^i_q(x^-_{i+\frac{1}{2}})}_{F^i_{qk}}$$

$$= (Eu^{i+1})_q - (Fu^i)_q,$$

where:

$$\mathbf{E^i_{qk}} = \frac{(-\mathbf{1})^{\mathbf{k}}}{\mathbf{2^{k+q}}} \quad and, \quad \mathbf{F^i_{qk}} = (\frac{\mathbf{1}}{\mathbf{2}})^{\mathbf{k+q}} \tag{4.20}$$

In conclusion, let $u^{n,i} = (u^{n,i}_0, \cdots, u^{n,i}_k, \cdots, u^{n,i}_K)^t$ be the following vector of $\mathbb{R}^{K+1}$ the schema (4.8) is given by :for $i = 1, ..., I$,

If a > 0,

$$M\frac{u^{n+1,i} - u^{n,i}}{\tau} + a(S+P)u^{n,i} - aQu^{n,i-1} = 0$$

equivalent to,

$$\frac{u^{n+1,i} - u^{n,i}}{\tau} + Au^{n,i} + Bu^{n,i-1} = 0 \tag{4.21}$$

where $\quad A = aM^{-1}(S+P), \quad$ and $\quad B = -aM^{-1}Q$

If a < 0, in the same way, we obtain:

$$\frac{u^{n+1,i} - u^{n,i}}{\tau} + D_1u^{n,i} + D_2u^{n,i+1} = 0 \tag{4.22}$$

where $\quad D_1 = aM^{-1}(S-F), \quad$ and $\quad D_2 = aM^{-1}E.$

4.1.2.2 Discontinuous Galerkin Scheme for the variable function a(x)

For the case where $a(x)$ is a smooth function, we use the scheme obtained in (4.6), and Taking $v = \varphi_q^i$. Interchanging the u by its polynomial approximation u_h leads to K+1 semi-discret equations for the same number of unknowns

$$\sum_{k=0}^{K}(\underbrace{\int_{C_i}\varphi_k^i(x)\varphi_q^i(x)dx}_{M_{qk}^i})\frac{du_k^i(t)}{dt}+\sum_{k=0}^{K}u_k^i(t)(\underbrace{\int_{C_i}a(x)\partial_x(\varphi_k^i(x))\varphi_q^i(x)dx}_{S_{qk}^i})+ \tag{4.23}$$

$$\underbrace{\frac{1}{2}(a(x_{i+\frac{1}{2}})-\mid a(x_{i+\frac{1}{2}})\mid)[u_h]_{i+\frac{1}{2}}(v_h)^-_{i+\frac{1}{2}}}_{(a)}+\underbrace{\frac{1}{2}(a(x_{i-\frac{1}{2}})+\mid a(x_{i-\frac{1}{2}})\mid)[u_h]_{i-\frac{1}{2}}(v_h)^+_{i-\frac{1}{2}}}_{(b)}=0$$

Before going any further, let $U^i = (u_0^i, \ldots, u_k^i, \ldots, u_K^i)^t$ be the unknown vector and $M_{qk}^i = \int_{C_i}\varphi_k^i(x)\varphi_q^i(x)dx$ be the mass matrix coefficients. We notice that this mass matrix is locally defined on each cell and is time independent.

The function $\phi(x) = a(x)\partial_x(\varphi_k^i(x))$ being at each point of C_i, then the matrix S_{qk}^i can be evaluated with an adapted quadrature formula, which can be of Gauss-integration. Now, we calculate the terms (a) and (b) given in equation (4.23),

$$(a) \Rightarrow \underbrace{\frac{1}{2}(a(x_{i+\frac{1}{2}})-\mid a(x_{i+\frac{1}{2}})\mid)}_{\widehat{b}_{i+\frac{1}{2}}}(u(x^+_{i+\frac{1}{2}})-u(x^-_{i+\frac{1}{2}}))(v_h)^-_{i+\frac{1}{2}}$$

$$=\widehat{b}_{i+\frac{1}{2}}(\sum_{k=0}^{K}u_k^{i+1}(t)\underbrace{\varphi_k^{i+1}(x^+_{i+\frac{1}{2}})\varphi_q^i(x^-_{i+\frac{1}{2}})}_{E_{qk}^{i+1}}-\sum_{k=0}^{K}u_k^i(t)\underbrace{\varphi_k^i(x^-_{i+\frac{1}{2}})\varphi_q^i(x^-_{i+\frac{1}{2}})}_{F_{qk}^i})$$

$$=\widehat{b}_{i+\frac{1}{2}}\sum_{k=0}^{K}u_k^{i+1}(t)E_{qk}^{i+1}-\widehat{b}_{i+\frac{1}{2}}\sum_{k=0}^{K}u_k^i(t)F_{qk}^i$$

The Matrices E and F are given in (4.20).

$$(b) \Rightarrow \underbrace{\frac{1}{2}(a(x_{i-\frac{1}{2}})+\mid a(x_{i-\frac{1}{2}})\mid)}_{\widehat{b}_{i-\frac{1}{2}}}(u(x^+_{i-\frac{1}{2}})-u(x^-_{i-\frac{1}{2}}))(v_h)^+_{i-\frac{1}{2}}$$

$$=\widehat{b}_{i+\frac{1}{2}}(\sum_{k=0}^{K}u_k^i(t)\underbrace{\varphi_k^i(x^+_{i-\frac{1}{2}})\varphi_q^i(x^+_{i-\frac{1}{2}})}_{P_{qk}^i}-\sum_{k=0}^{K}u_k^{i-1}(t)\underbrace{\varphi_k^{i-1}(x^-_{i-\frac{1}{2}})\varphi_q^i(x^+_{i-\frac{1}{2}})}_{Q_{qk}^{i-1}})$$

$$= \widehat{b}_{i-\frac{1}{2}} \sum_{k=0}^{K} u_k^i(t) P_{qk}^i - \widehat{b}_{i-\frac{1}{2}} \sum_{k=0}^{K} u_k^{i-1}(t) Q_{qk}^{i-1}$$

The Matrices P and Q are given in (4.18).

Finally, the K+1 semi-discrete equation (4.1) can be written in the following compact form

$$M^i \frac{U^{i,n+1} - U^{i,n}}{\tau} + (S^i + \widehat{b}_{i-\frac{1}{2}} P^i - \widehat{b}_{i+\frac{1}{2}} F^i) U^{i,n} + \widehat{b}_{i+\frac{1}{2}} E^{i+1} U^{i+1,n} - \widehat{b}_{i-\frac{1}{2}} Q^{i-1} U^{i-1,n} = 0 \tag{4.24}$$

Then,

$$\frac{U^{i,n+1} - U^{i,n}}{\tau} + A^i U^{i,n} + B^{i+1} U^{i+1,n} + D^{i-1} U^{i-1,n} = 0 \tag{4.25}$$

where,

$$\begin{cases} A^i = M^{-1}(S^i + \widehat{b}_{i-\frac{1}{2}} P^i - \widehat{b}_{i+\frac{1}{2}} F^i) \\ B^i = M^{-1} \widehat{b}_{i+\frac{1}{2}} E^{i+1} \\ D^i = -M^{-1} \widehat{b}_{i-\frac{1}{2}} Q^{i-1}. \end{cases}$$

4.1.3 Projection onto the approximation space

Being given a function $\phi = \phi(x)$, we compute its polynomial approximation restricted to the cell C_i as $\phi_h^i(x) = \sum_{k=0}^{K} \phi_k^i \varphi_k^i$. To this end we define the projection operator $\prod$ as

$$\begin{aligned} \Pi : L^1(C_i) &\rightarrow \mathbb{P}^K(C_i) \\ \phi &\rightarrow \Pi(\phi) = \phi_h^i, \end{aligned} \tag{4.26}$$

where the polynomial approximation $\Pi(\phi)$is defined by its successive moments ϕ_k^i as

$$\int_{C_i} \Pi(\phi) \varphi_q^i dx = \int_{C_i} \phi \varphi_q^i dx, \qquad forq = 0, \ldots, K.$$

This projection operator is also used to construct the polynomial approximation of the initial data $u^0(x)$

$$u_h^{i,0} = \sum_{k=0}^{K} u_k^{i,0} \varphi_k(x),$$

where the $u_k^{i,0}$ coefficients are obtained by solving the linear system

$$MU^{i,0} = \widetilde{U}^{i,0}.$$

$\widetilde{U}^{i,0} = (\langle u^0, \varphi_0^i \rangle, \ldots, \langle u^0, \varphi_K^i \rangle)^t$,

We have, for $q = 0, \dots, K$,

$$\int_{C_i} u(x,0)\varphi_q^i dx = \int_{C_i} u_0(x)\varphi_q^i(x)dx,$$

$$= \sum_{k=0}^{K} u_k^i(0) \underbrace{\int_{C_i} \varphi_k^i(x)\varphi_q^i(x)}_{M_{qk}}$$

then,

$$\int_{C_i} u_0(x)\varphi_q^i dx = (M)_q U^{i,0}, \qquad (a)$$

on the other hand we have: for $q = 0, \dots, K$,, we make a variable change

$$\int_{C_i} u_0(x)\varphi_q^i dx = \int_{-1}^{1} u_0(\frac{h}{2}(y+1) + x_{i-\frac{1}{2}})\varphi_q^i(\frac{h}{2}(y+1) + x_{i-\frac{1}{2}})\frac{h}{2}dy,$$

The Gauss integration method is applied, we have:(K_g the number of the gauss points)

$$\int_{C_i} \underbrace{u_0(x)\varphi_q^i(x)}_{f(x)} dx = \sum_{p=0}^{K_g} w_p^i f(x_p^i),$$

where, $w_p^i = \frac{h}{2}w_p$; w_p is the of weights Gauss, and $x_p^i = x_{i-\frac{1}{2}} + \frac{h}{2}(x_p+1)$; x_p is the Gauss points,
We define the vector

$$\widetilde{U}^{i,0} = (\int_{C_i} u_0(x)\varphi_0^i dx, \dots, \int_{C_i} u_0(x)\varphi_q^i dx, \dots, \int_{C_i} u_0(x)\varphi_K^i dx)^t \qquad (b)$$

According to (a) and (b), we obtain:

$$MU^{i,0} = \widetilde{U}^{i,0} \qquad \Rightarrow \qquad U^{i,0} = M^{-1}\widetilde{U}^{i,0}.$$

4.1.4 L_2 Stability of the DG-scheme

In this part, following [116, 32, 84], we want to design a generic form of numerical fluxes which ensures the global stability of our semi-discrete discretization in the L_2 norm. To this end, we integrate by parts the equation (4.6), as follows: for any $v_h \in V_h^k$,

$$\int_{C_i} \frac{\partial u_h}{\partial t}(x,t)v_h(x)dx - \int_{C_i} a(x)u_h(x,t)\partial_x v_h(x)dx + \widehat{au_h}_{i+\frac{1}{2}}(v_h)^-_{i+\frac{1}{2}} - \widehat{au_h}_{i-\frac{1}{2}}(v_h)^+_{i-\frac{1}{2}}$$
$$= \int_{C_i} a_x(x)u_h(x,t)v_h(x)dx \qquad (4.27)$$

where

$$\widehat{au_h}_{i+\frac{1}{2}} = \frac{1}{2}(a(x_{i+\frac{1}{2}})+\mid a(x_{i+\frac{1}{2}})\mid)(u_h)^-_{i+\frac{1}{2}} + \frac{1}{2}(a(x_{i+\frac{1}{2}})-\mid a(x_{i+\frac{1}{2}})\mid)(u_h)^+_{i+\frac{1}{2}}$$

Then, we obtain the standard DG scheme for conservation laws equation with source terms using Roe fluxes [45]. We can, therefore, use the standard techniques in the analysis for the DG schemes to obtain the following theoretical results. In this part, following [32, 84], we want study the stability of semi-discrete discretization on the L_2 norm.

Proposition 1.

The solution u_h to the semi-discrete DDG scheme (4.27) satisfies the following cell entropy inequality

$$\frac{d}{dt}\int_{C_i} U(u_h)dx + \widehat{F}_{i+\frac{1}{2}} - \widehat{F}_{i-\frac{1}{2}} - \int_{C_i} a_x(u_h)^2dx \leq 0 \tag{4.28}$$

for the square entropy $U(u) = \frac{u^2}{2}$, for some consistent entropy flux

$$\widehat{F}_{i+\frac{1}{2}} = \widehat{F}(u_h(x^-_{i+\frac{1}{2}}, t), u_h(x^+_{i+\frac{1}{2}}, t))$$

satisfying $\widehat{F}(u, u) = F(u)$

Proof. Let us consider the local variational formulation written using u_h^i as a test function

$$\int_{C_i} \frac{\partial u_h}{\partial t}(x,t)u_h dx - \int_{C_i} a(x)u_h\partial_x v_h dx + \widehat{au_h}_{i+\frac{1}{2}}(u_h)^-_{i+\frac{1}{2}} - \widehat{au_h}_{i-\frac{1}{2}}(u_h)^+_{i-\frac{1}{2}}$$
$$= \int_{C_i} a_x(x)u_h(x,t)u_h(x)dx$$

We can immediately infer

$$\frac{1}{2}\frac{d}{dt}\int_{C_i}(u_h)^2dx - \int_{C_i} a(x)u_h\partial_x v_h dx + \widehat{au_h}_{i+\frac{1}{2}}(u_h)^-_{i+\frac{1}{2}} - \widehat{au_h}_{i-\frac{1}{2}}(u_h)^+_{i-\frac{1}{2}}$$
$$= \int_{C_i} a_x(x)(u_h)^2dx \tag{4.29}$$

Here, we make use of the function F which denotes a primitive of the flux function defined as $F(u) = \int^u f(s)ds$. Noticing that $\frac{\partial F(u_h)}{\partial x} = \frac{dF(u_h)}{du}\frac{\partial u_h}{\partial x} = f(u_h)\frac{\partial u_h}{\partial x}$, where $f(u_h) = a(x)u_h$, equation (4.29) writes

$$\frac{1}{2}\frac{d}{dt}\int_{C_i}(u_h)^2dx - [F(u_h)]_{x_{i-\frac{1}{2}}}^{x_{i+\frac{1}{2}}} + \widehat{f}_{i+\frac{1}{2}}(u_h)^-_{i+\frac{1}{2}} - \widehat{f}_{i-\frac{1}{2}}(u_h)^+_{i-\frac{1}{2}}$$
$$= \int_{C_i} a_x(x)(u_h)^2dx \tag{4.30}$$

Let us set $R_i = \widehat{f}_{i+\frac{1}{2}}(u_h)^-_{i+\frac{1}{2}} - \widehat{f}_{i-\frac{1}{2}}(u_h)^+_{i-\frac{1}{2}} - [F(u_h)]^{x_{i+\frac{1}{2}}}_{x_{i-\frac{1}{2}}}$, then:

$$\frac{1}{2}\frac{d}{dt}\int_{C_i}(u_h)^2dx + R_i - \int_{C_i} a_x(x)(u_h)^2dx = 0 \tag{4.31}$$

The other hand we have:

$$R_i = -F^-_{i+\frac{1}{2}} + \widehat{f}_{i+\frac{1}{2}}(u_h)^-_{i+\frac{1}{2}} + F^+_{i-\frac{1}{2}} - \widehat{f}_{i-\frac{1}{2}}(u_h)^+_{i-\frac{1}{2}},$$

$$R_i = \widehat{F}_{i+\frac{1}{2}} - \widehat{F}_{i-\frac{1}{2}} + \Theta_i \tag{4.32}$$

where, $\begin{cases} \widehat{F}_{i+\frac{1}{2}} = -F^-_{i+\frac{1}{2}} + \widehat{f}_{i+\frac{1}{2}}(u_h)^-_{i+\frac{1}{2}} \\ \widehat{F}_{i-\frac{1}{2}} = -F^-_{i-\frac{1}{2}} + \widehat{f}_{i-\frac{1}{2}}(u_h)^-_{i-\frac{1}{2}} \end{cases}$

and $\Theta_i = -F^-_{i-\frac{1}{2}} + \widehat{f}_{i-\frac{1}{2}}(u_h)^-_{i-\frac{1}{2}} + F^+_{i-\frac{1}{2}} - \widehat{f}_{i-\frac{1}{2}}(u_h)^+_{i-\frac{1}{2}}$

$$\Theta_i = F^+_{i-\frac{1}{2}} - F^-_{i-\frac{1}{2}} - \widehat{f}_{i-\frac{1}{2}}((u_h)^+_{i-\frac{1}{2}} - (u_h)^-_{i-\frac{1}{2}}) \tag{4.33}$$

Remark. The physically relevant weak solution of conservation laws equation (the so-called entropy solution) satisfies the following entropy inequality:

$$U(u)_t + F(u)_x \leq 0.$$

According to the remark, we must show that $\Theta_i \geq 0$, we applied the mean value theorem between $(u_h)^+_{i-\frac{1}{2}}$ and $(u_h)^-_{i-\frac{1}{2}}$, then,

$\exists \xi \in](\mathbf{u_h})^-_{\mathbf{i-\frac{1}{2}}}, (\mathbf{u_h})^+_{\mathbf{i-\frac{1}{2}}}[$ such as

$$\mathbf{F}^+_{\mathbf{i-\frac{1}{2}}} - \mathbf{F}^-_{\mathbf{i-\frac{1}{2}}} = \mathbf{F}'(\xi)((\mathbf{u_h})^+_{\mathbf{i-\frac{1}{2}}} - (\mathbf{u_h})^-_{\mathbf{i-\frac{1}{2}}}) \tag{4.34}$$

Now, one replaces (4.34) in formula (4.33) we have

$$\Theta_i = -(\widehat{f}_{i-\frac{1}{2}} - F'(\xi))((u_h)^+_{i-\frac{1}{2}} - (u_h)^-_{i-\frac{1}{2}}) \geq 0 \tag{4.35}$$

Under the definition of E-flux given in [84]

$$(h(u^-, u+) - f(u))(u^+ - u^-) \leq 0$$

Thus,

$$\frac{1}{2}\frac{d}{dt}\int_{C_i}(u_h)^2dx + \widehat{F}_{i+\frac{1}{2}} - \widehat{F}_{i-\frac{1}{2}} - \int_{C_i} a_x(x)(u_h)^2dx \leq 0 \tag{4.36}$$

□

Proposition 2.

Suppose there is a constant β such that the derivative of $a(x)$ satisfies $\partial_x a(x) < \beta$ for $x \in [x_{min}, x_{max}]$, then we have the following L^2 Stability for the scheme(4.23):

$$\|u_h(t)\|_{L^2} \leq e^{\beta t}\|u_h(0)\|_{L^2}$$

Proof. Using the result given in the previous proposition 4, we have for periodic or compact boundary conditions, the sum of (4.36) over all the cells writes

$$\frac{1}{2}\frac{d}{dt}\|u_h\|_{L_2}^2 dx = \frac{1}{2}\frac{d}{dt}\sum_{i,cells}\int_{C_i}(u_h)^2 dx \leq \sum_{i,cells}\int_{C_i} a_x(x)(u_h)^2 dx \tag{4.37}$$

Such as $a_x < \beta$, we have

$$\frac{1}{2}\frac{d}{dt}\int_{x_{min}}^{x_{max}}(u_h)^2 dx \leq \beta\int_{x_{min}}^{x_{max}}(u_h)^2 dx$$

Integrating over t, one obtains:

$$\int_0^t \left(\frac{\frac{d}{dt}\int_{x_{min}}^{x_{max}}(u_h)^2 dx}{\int_{x_{min}}^{x_{max}}(u_h)^2 dx}\right)dt \leq 2\beta\int_0^t dt.$$

Then,

$$\|u_h(t)\|_{L^2} \leq e^{\beta t}\|u_h(0)\|_{L^2}\ .$$

□

4.1.5 Error Estimates for Smooth Solutions

If we assume the exact solution of (4.1) is smooth where $a(x) = cst$, we can obtain L^2-error estimates. Such error estimates can be obtained for the general nonlinear conservation law and for fully discretized RKDG methods, see the lecture book of cockburn[42]. In this part,for simplicity we will give only the proof for the semi-discret DDG scheme and the linear version of (4.1).

Proposition 3.

If the function $a(x)$ and the solution u of the equation (4.1) are smooth and the DDG scheme (4.6) with the finite-dimensional function space V_h^K is used, then we have the following optimal L^2 error estimate

$$\| u_h(t) - u(t) \|_{L^2} \leq C(\Delta x)^{K+1} \tag{4.38}$$

where C depends on u and its derivatives but is independent of Δx.

Proof. For the DDG scheme (4.6), we introduce a short-hand notation:

$$B_i(u_h; v_h) = \int_{C_i} (\frac{\partial u_h}{\partial t}(x,t) + a(x)\frac{\partial u_h}{\partial x}(x,t))v_h(x)dx + \frac{1}{2}(a(x_{i+\frac{1}{2}}) - \mid a(x_{i+\frac{1}{2}}) \mid)[u_h]_{i+\frac{1}{2}}(v_h)^-_{i+\frac{1}{2}}$$
$$+ \frac{1}{2}(a(x_{i-\frac{1}{2}}) + \mid a(x_{i-\frac{1}{2}}) \mid)[u_h]_{i-\frac{1}{2}}(v_h)^+_{i-\frac{1}{2}} \tag{4.39}$$

Then, the equation (4.6) can be written as:

$$B_i(u_h; v_h) = 0, \tag{4.40}$$

for all $v_h \in V_h^K$ and for all i. and the exact solution of the PDE (4.1) also satisfies

$$B_i(u; v_h) = 0, \tag{4.41}$$

for all $v_h \in V_h^K$ and for all i. Subtracting (4.40) from (4.40) and using the linearity of B_i with respect to its first argument, we obtain the error equation

$$B_i(u - u_h; v_h) = 0, \tag{4.42}$$

for all $v_h \in V_h^K$ and for all i.
We now define a spacial projection P into V_h^K . For a given smooth function w, the projection Pw is the unique function in V_h^K which satisfies, for each i,

$$\int_{C_i} (Pw(x) - w(x))v_h(x)dx = 0 \qquad \forall v_h \in P^{K-1}(C_i); \tag{4.43}$$

Standard approximation theory of Ciarelet [41] implies, for a smooth function w,

$$\parallel Pw(x) - w(x) \parallel \leq C(\Delta x)^{K+1} \tag{4.44}$$

where here and below C is a generic constant depending on w and its derivatives but independent of Δx. We take:

$$v_h = Pu - u_h$$

in the error equation (4.42), and denote

$$e_h = Pu - u_h, \qquad \varepsilon_h = u - Pu \tag{4.45}$$

to obtain

$$B_i(e_h; e_h) = -B_i(\varepsilon_h; e_h). \tag{4.46}$$

For the left-hand side of (4.46), we use the cell entropy inequality given in (4.36) to obtain

$$B_i(e_h; e_h) = \frac{1}{2}\frac{d}{dt}\int_{C_i} (e_h)^2 dx + \widehat{F}_{i+\frac{1}{2}} - \widehat{F}_{i-\frac{1}{2}} - \int_{C_i} a_x(x)(e_h)^2 dx \tag{4.47}$$

Using the properties (4.43) of the projection P for the right-hand side of (4.46), we have

$$-B_i(\varepsilon_h; e_h) = -\int_{C_i} (\varepsilon_h)_t e_h dx + \int_{C_i} a_x(x)(\varepsilon_h) e_h dx \qquad (4.48)$$

Therefore, $B_i(\varepsilon_h; e_h)$ becomes

$$-B_i(\varepsilon_h; e_h) \leq \frac{1}{2}(\int_{C_i} ((\varepsilon_h)_t)^2 dx + \beta \int_{C_i} (e_h)^2 dx) \qquad (4.49)$$

Summing up over i the equation (4.48) and (4.49), and using the approximation result (4.44), we have

$$\frac{d}{dt}\int_{x_{min}}^{x_{max}} (e_h)^2 dx \leq \int_{x_{min}}^{x_{max}} (e_h)^2 dx + C(\Delta x)^{2K+2}$$

A Gronwall's inequality, the fact that initial error

$$\| u(.,0) - u_h(.,0) \leq C(\Delta x)^{K+1} \|$$

The initial condition $u_h(.,0)$ is taken as the L^2-projection of the analytical initial condition $u(.,0)$, and the approximation result (4.44) finally give us the error estimate (4.38). □

4.1.6 The TVD-Runge-Kutta time discretization

To discretize our ODE system in time, we use the TVD Runge Kutta time discretization introduced by Cockburn & Shu (1989) .We can write equation (12) as:

$$\frac{u^{n+1} - u^n}{\tau} = L(u^n) \qquad (4.50)$$

Where,

$$L(u^n) = -(Au^n + Bu^n)$$

The algorithm for Runge-Kutta method is given as follow:
The method is more accurate and more the stability condition is binding. For the discontinuous Galerkin scheme with a Runge-Kutta methods, the CFL coefficient is of the form:

$CFL = \frac{1}{2k-1}$, with k is the order of our method

4.1.7 Precision of the DG-scheme

To assess the accuracy of a numerical scheme, we evaluate the difference between the numerical solution and the analytical solution employing a sequence of successively refined grids characterized by a uniform mesh spacing $h = \Delta x$. Let u_h^n be the numerical solution

Algorithm 6 Algorithm for Runge-Kutta Method

Require: Thus, if $(t^n)_{n=0}^{n=N}$ is a partition of [0,T] and $\tau^n = t^{n+1} - t^n; n = 0, ..., N-1$;
Our time-marching algorithm reads as follows:
We will use the following standard RK methods for a system of the form:
if The first method is chosen **then**
return $u^{n+1} = u^n + \tau L u^n \ \equiv: \ S(t_n, u^n)$ (Euler scheme)
else
if The second method is chosen **then**
return $u^{n+1} = \frac{1}{2}(u^n + S(t_{n+1}, S(t_n, u^n)))$ (TVD-RK2 scheme)
else {The third method is chosen}
return (TVD-RK3 scheme) with $t_{n+\frac{1}{2}} = t_n + \frac{\tau}{2}$:

$$u^{n,1} := \frac{1}{2}(u^n + S(t_{n+1}, S(t_n, u^n))), \tag{4.51a}$$

$$u^{n,2} := \frac{3}{4}u^n + \frac{1}{4}u^{n,1}, \tag{4.51b}$$

$$u^{n+1} := \frac{1}{3}u^n + \frac{2}{3}S(t_{n+\frac{1}{2}}, u^{n,2}). \tag{4.51c}$$

Set $u^0(x) = u_0(x)$
for $n = 0$ to $N-1$ **do**
compute u^{n+1} from u^n as follows :
Set $u^0 = u^n$;
for $i = 1$ to $k+1$ **do**
compute the intermediate functions:

$$u^i = \{\sum_{k=0}^{i-1} \alpha_{il} u^l + \beta_{il} \tau L(u^{(l)})\};$$

where α and β is the parameters of Runge-Kutta Method
set u^{n+1}=u^n

obtained from the numerical scheme at time tn, i:e:, $u_h^n(x) = u_i^n$ for $x \in Ci$. Denoting by $u_{ex} = u_{ex}(x; t^n)$ the analytical solution, we define the numerical error at time t^n as:

$$E_{L_\alpha} = \|u_h^n - u_{ex}\|_{L_\alpha},$$

where $\|.\|_{L_\alpha}$ is a given norm characterized by the subscript α. We will assess the numerical errors for the three following classical norms: L_1, L_2, L_∞. For theses norms the corresponding numerical errors read

$$E_{L_1} = \sum_{i \in C_i} |u_i^n - u_{ex}(x, t^n)| dx$$

$$E_{L_2} = \sqrt{\sum_{i \in C_i} [u_i^n - u_{ex}(x, t^n)]^2 dx}$$

$$E_{L_\infty} = \max_i [\max_{x \in C_i} (|u_i^n - u_{ex}(x, t^n)|)]$$

For a grid characterized by the constant mesh spacing Δx, we suppose that the numerical error E_{L_α} has the following behavior

$$E_{L_\alpha}^{\Delta x} = D(\Delta x)^{q_{L_\alpha}}, \tag{4.52}$$

where D is non-negative constant dependent of Δt, but it's independent of Δx and q_{L_α} is a real which corresponds to the rate of convergence of the numerical schema. To compute the rate of convergence, we evaluate the numerical errors corresponding to two grids characterized by Δx_1 and Δx_2 with $\Delta x_1 > \Delta x_2$. Using the relation (4.52) yields

$$q_{L_\alpha}^{\Delta x} = \frac{\log(E_{L_\alpha}^{\Delta x_1}) - \log(E_{L_\alpha}^{\Delta x_2})}{\log(\Delta x_1) - \log(\Delta x_2)}. \tag{4.53}$$

4.1.8 Computational Results

Example 1.

We provide preliminary numerical results for the schemes discussed in this section. We solve the transport's equation with a periodic boundary condition:

$$\begin{cases} u_t + a u_x = 0, & \forall (t, x) \in [0, 1] \times [0, 1], \\ u_0(x) = sin(2\pi x), & \forall x \in [0, 1]. \end{cases}$$

The exact solution is $u(x, t) = sin(2\pi(x - at))$

Results for the L^∞ error and we use $RK2$, further corresponding orders are given in Table 1.

	K=1		K=2		K=3	
Mesh I	error	order	error	order	error	order
20	0.1872E-03	-	0.1274E-03	-	0.4875E-04	-
40	0.5600E-04	1.78	0.1910E-04	2.73	0.3452E-05	3.81
80	0.1709E-04	1.79	0.2510E-05	2.98	0.2173E-06	3.98
160	0.4389E-05	1.97	0.3174E-06	2.99	0.1285E-07	4.07

Table 4.1: L^∞errors, advection test with $u_0(x) = sin(2\pi x)$

In Table 2, we check time accuracy for fixed Δ x, we study the orders of the DG-RKp schemes when $p \in 1, 2, 3$.

	Euler		RK2		RK3	
Mesh N	error	order	error	order	error	order
20	0.9868E-04	-	0.1024E-06	-	0.2263E-06	-
40	0.4934E-04	0.99	0.2495E-07	2.03	0.3109E-07	2.86
80	0.2476E-04	1.00	0.6172E-08	2.01	0.4050E-08	2.94
160	0.6167E-05	1.00	0.1549E-08	1.99	0.5603E-09	2.85

Table 4.2: Example 1 (1D advection) with fixed spatial order k=4 and mesh m=100, to check time accuracy

Example 2.
We consider now the case of an advection equation with nonconstant advection term, we suppose the function a(x) is positive.

$$\begin{cases} u_t + (sin(2\pi x) + d)u_x = 0, & d > 1 \quad \forall (t,x) \in [0,1] \times [0,1], \\ u_0(x) = sin(2\pi x), & \forall x \in [0,1]. \\ u(0,t) = u(2\pi, t) & \forall t \in [0,1]. \end{cases}$$

we use the characteristic method to calculate The exact solution, as follows :

$$u(x,t) = sin(2\arctan[\frac{d_1}{d}\tan(\arctan(y(x)) - d_1\pi t) - \frac{1}{d}])$$

where, $d_1 = \sqrt{d^2-1}$, and $y(x) = \frac{d\tan(\pi x)+1}{d_1}$
Results are given in Table 3, we take $d = 1.5$ and the terminal time $T = 0.5$

	K=0		K=1		K=2	
Mesh I	error	order	error	order	error	order
40	0.58E-01	-	0.64E-03	-	0.32E-02	-
80	0.36E-01	0.96	0.05E-04	1.92	0.31E-03	2.91
160	0.24E-01	0.97	0.46E-05	1.96	0.34E-04	2.94
320	0.95E-02	0.99	0.89E-06	1.98	0.37E-05	2.97

Table 4.3: L^2errors, advection test non-constant advection term with initial $u_0(x) = sin(2\pi x)$

Example 3.
The last example for advection equation with nonconstant advection term,

$$\begin{cases} u_t + sin(2\pi x)u_x = 0, & \forall (t,x) \in [0,1] \times [0,1], \\ u_0(x) = sin(2\pi x), & \forall x \in [0,1]. \\ u(0,t) = u(2\pi, t) & \forall t \in [0,1]. \end{cases}$$

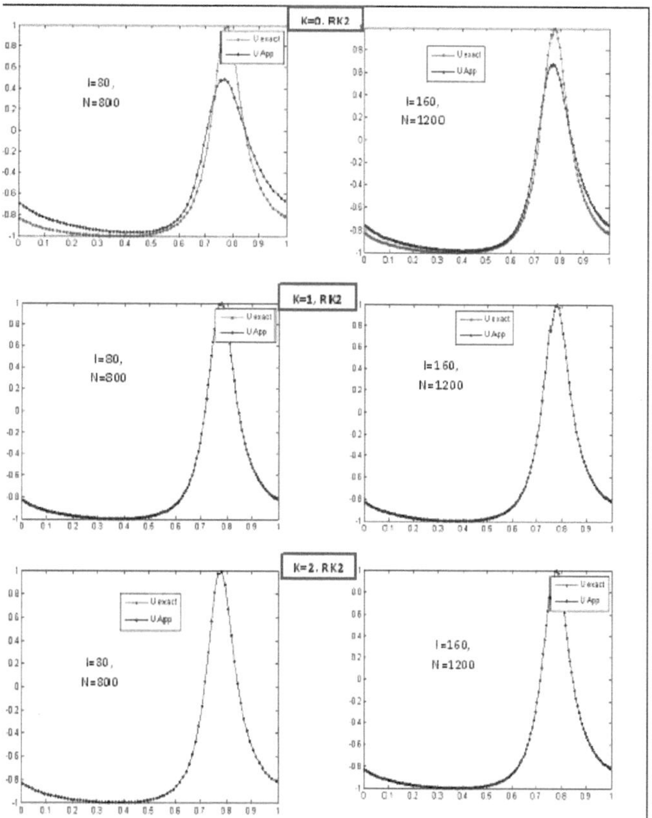

Figure 4.1: Advection test non-constant advection term with initial $u_0(x) = sin(2\pi x)$

we use the characteristic method to calculate The exact solution, as follows :

$$u(x,t) = sin(2\arctan[e^{-2\pi t}.\tan(\pi x)])$$

Results are given in Table 4, we take the terminal time $T = 0.5$ and $CFL = 0.2$
Example 4. The Burgers' equation

$$\begin{cases} u_t + \frac{u_x^2}{2} = 0, \quad d > 1 & \forall (t,x) \in [0,2\pi] \times [0,1], \\ u_0(x) = \sin(x), & \forall x \in [0,2\pi]. \\ u(0,t) = u(2\pi,t) & \forall t \in [0,t]. \end{cases}$$

	K=1		K=2	
Mesh I	error	order	error	order
40	0.92E-04	-	0.158E-04	-
80	0.242E-05	1.93	0.229E-05	2.79
160	0.621E-06	1.98	0.323E-06	2.82
320	0.148E-06	2.00	0.423E-07	2.94

Table 4.4: L^2-Errors, advection test non-constant advection term with initial $u_0(x) = sin(2\pi x)$

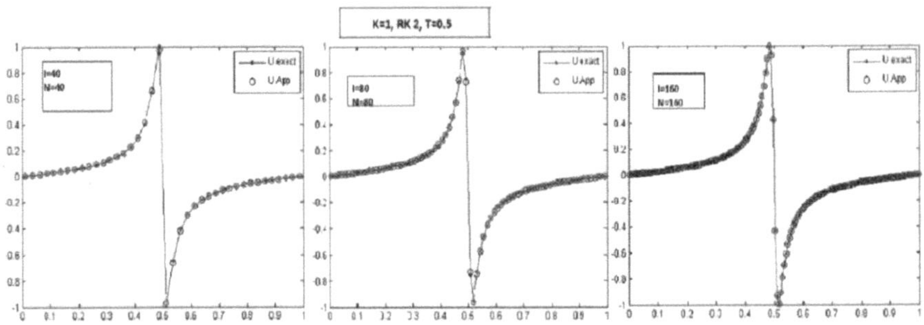

Figure 4.2: Advection test non-constant advection term with initial $u_0(x) = sin(2\pi x)$

The exact solution when ϕ is still smooth is obtained by the characteristics methods. First solve x_0 from $x = x_0 + \cos(x_0)t$. The errors and numerical orders of accuracy when using P^2 polynomials and the third Runge-Kutta for the time discretization are given in the table as follow:

Mesh N	L^2 error	order	L^∞ error	order
40	0.69E-03	-	0.52E-02	-
80	0.18E-03	1.93	0.14E-02	1.89
160	0.45E-04	1.98	0.36E-03	1.96
320	0.11E-04	2.00	0.89E-04	2.00

Table 4.5: L^2-Errors,and L^∞-Errors,for the burgers equation with initial $u_0(x) = \sin(x)$ and final time t=1,CFL=0.08

4.2 The RKDDG method for pricing European Option

In this section we describe briefly the Runge Kutta Direct Discontinuous Galerkin (RKDDG) Method for the European option problem.
We consider the PDE of the European put option with Strike $K > 0$ and maturity $T > 0$:

$$v_t - \frac{\sigma^2}{2}s^2 v_{ss} - rsv_s + rv = 0, \quad t \in (0,T), \ s \in \mathcal{D} \tag{4.54a}$$
$$v(0,x) := \max(K - s, 0), \quad s \in \mathcal{D} \tag{4.54b}$$

with $\mathcal{D} = (S_{min}, S_{max})$, where $0 < S_{min} < K < S_{max}$ and S_{max} is large enough. In the first, considering the new functions $a(x)$, $b(x)$, and $c(x)$, the problem (4.54) becomes

$$u_t - a(x)u_{xx} + b(x)u_x + c(x)u = 0, \quad t \in (0,T), \ x \in \Omega \tag{4.55a}$$
$$u(0,x) := \max(K - x, 0), \quad x \in \Omega \tag{4.55b}$$

where, $a(x) = \frac{\sigma^2}{2}s^2$, $b(x) = -rs$, $c(x) = r$
We will formulate a (DDG) scheme

4.2.1 The Scheme for the Second order diffusion problem

We now turn our attention to the convection diffusion problems containing second derivatives. The difficulty can be illustrated by the following simple heat equation:

In this section; $a > 0$ represent a diffusion coefficient. (We denote $\Omega = [x_{min}, x_{max}]$)

$$\begin{aligned} \partial_t u(x,t) - a\partial_{xx}u(x,t) &= 0 \qquad (x,t) \in \Omega \times [0,T], & (4.56a)\\ u(x,0) &= u_0(x) & (4.56b) \end{aligned}$$

with periodic boundary conditions. After introducing the new variable $r(x,t)$ the heat equation (4.56) as a first order system:

$$\begin{aligned} \partial_t u(x,t) + a\partial_x r(x,t) &= 0 \qquad (x,t) \in \Omega \times [0,T], & (4.57a)\\ r(x,t) + \partial_x u(x,t) &= 0 \qquad (x,t) \in \Omega \times [0,T], & (4.57b)\\ u(x,0) &= u_0(x) & (4.57c) \end{aligned}$$

We can then formally use the same discontinuous Galerkin method as in section 1 for the convection equation to solve(4.52), resulting in the following scheme: find $u, r \in V_h$ such that,

$$\int_{C_i} \partial_t u_h(x,t) v_h(x) dx + a \int_{C_i} \partial_x r_h(x,t) v_h(x) + a[r]_{i-\frac{1}{2}} v(x^+_{i-\frac{1}{2}}) = 0 \quad \forall (x,t) \in \Omega \times [0,T], \tag{4.58}$$

$$\int_{C_i} r_h(x,t) w_h(x) dx + \int_{C_i} \partial_x u_h(x,t) w_h(x) + [u]_{i+\frac{1}{2}} w(x^-_{i+\frac{1}{2}}) = 0 \quad \forall (x,t) \in \Omega \times [0,T]. \tag{4.59}$$

For all $v_h, w_h \in V_h$ and, $i = 1, ..., I$.

A good choice for the numerical fluxes, both in accuracy and in compactness of the eventual stencil after the auxiliary variable r_h is eliminated, is $\widehat{u}_{j+\frac{1}{2}} = u_h(x^+_{j+\frac{1}{2}}, t)$, $\widehat{r}_{j+\frac{1}{2}} = r_h(x^-_{j+\frac{1}{2}}, t)$ We decompose u_h and r_h on the basis V_h and replace in the equation:

$$u_h(x,t) = \sum_{k=0,..,K} u_k^{n,i} \varphi_k^i, \tag{4.60a}$$

$$r_h(x,t) = \sum_{k=0,..,K} r_k^{n,i} \varphi_k^i \tag{4.60b}$$

Then,

$$\sum_{k=0,..,K} \partial_t u_k^i(t) \int_{C_i} \varphi_k^i(x) \varphi_q(x) dx + a \sum_{k=0,..,K} r_k^i(t) \int_{C_i} \partial_x \varphi_k^i(x) \varphi_q^i(x) dx + a[r]_{i-\frac{1}{2}} \varphi_q(x^+_{i-\frac{1}{2}}) = 0 \tag{4.61}$$

$$\sum_{k=0,..,K} r_k^i(t) \int_{C_i} \varphi_k^i(x) \varphi_q(x) dx + \sum_{k=0,..,K} u_k^i(t)) \int_{C_i} \partial_x \varphi_k^i(x) \varphi_q^i(x) dx + [u]_{i+\frac{1}{2}} \varphi_q^i(x^-_{i+\frac{1}{2}}) = 0 \tag{4.62}$$

Using the results of the Section(2.1), we calculate the matrices associated to equation (4.61): We denote,

$$M_{qk} = \int_{C_i} \varphi_k(x) \varphi_q(x) dx, \quad and \quad S_{qk} = \int_{C_i} \partial_x \varphi_k(x) \varphi_q(x) dx$$

- $[r]_{i-\frac{1}{2}} \varphi_q(x^+_{i-\frac{1}{2}}) = Pr^i - Qr^{i-1}$,
  where P and Q are calculated in (4.18).

- $[u]_{i+\frac{1}{2}} \varphi_q(x^-_{i+\frac{1}{2}}) = Eu^{i+1} - Fu^i$,
  where E and F are calculated in (4.20).

The Euler Scheme (Forward) for the heat equation is given by :

$$\begin{cases} \dfrac{u^{n+1,i} - u^{n,i}}{\tau} + Ar^{n,i} + Br^{n,i-1} = 0, \\ r^{n,i} + D_1 u^{n,i} + D_2 u^{n,i+1} = 0 \end{cases} \tag{4.63}$$

where, A and B are as in (4.21):

$$A = aM^{-1}(S+P), \qquad B = -aM^{-1}Q$$

and D_1 and D_2 are computed in the same way as in (4.22)

$$D_1 = M^{-1}(S-F), \qquad D_2 = M^{-1}E$$

We use $RK2$ and $RK3$ for the equation (4.63) (see appendix A).

4.2.2 The Scheme for the Convection-Diffusion problem

In this section, we present the LDG methods for the following simple model problem:

$$\partial_t u(x,t) - a(x)\partial_{xx}u(x,t) + b(x)\partial_x u(x,t) + c(x)u(x,t) = 0 \qquad (x,t) \in \Omega \times [0,T], \tag{4.64a}$$

$$u(x,0) = u_0(x) \tag{4.64b}$$

To define the LDG method, we introduce the new variable $q(x,t) = \partial_x u(x,t)$ and rewrite the problem (4.64) as follows:

$$\partial_t u(x,t) + a(x)\partial_x q(x,t) + b(x)u(x,t) + c(x)u(x,t) = 0 \qquad (x,t) \in \Omega \times [0,T], \tag{4.65a}$$

$$q(x,t) + \partial_x u(x,t) = 0 \qquad (x,t) \in \Omega \times [0,T], \tag{4.65b}$$

$$u(x,0) = u_0(x) \tag{4.65c}$$

The semi-discrete LDDG scheme of (4.65) is defined as follows:
Find u_h, $q_h \in V_h^k$, such as for any test function $v_h, w_h \in V_h^k$, and $\forall 1 \leq i \leq N$, we have:

$$\int_{C_i}(\frac{\partial u_h}{\partial t}(x,t)+a(x)\frac{\partial q_h}{\partial x}(x,t)+b(x)\frac{\partial u_h}{\partial x}(x,t)+c(x)u_h)v_h(x)dx+a(x_{i-\frac{1}{2}})[q_h]_{i-\frac{1}{2}}(v_h)^+_{i-\frac{1}{2}}$$
$$+\frac{1}{2}(b(x_{i+\frac{1}{2}})-\mid b(x_{i+\frac{1}{2}})\mid)[u_h]_{i+\frac{1}{2}}(v_h)^-_{i+\frac{1}{2}}+\frac{1}{2}(b(x_{i-\frac{1}{2}})+\mid b(x_{i-\frac{1}{2}})\mid)[u_h]_{i-\frac{1}{2}}(v_h)^+_{i-\frac{1}{2}} \quad (4.66)$$
$$+\frac{1}{2}(c(x_{i+\frac{1}{2}})-\mid c(x_{i+\frac{1}{2}})\mid)[u_h]_{i+\frac{1}{2}}(v_h)^-_{i+\frac{1}{2}}+\frac{1}{2}(c(x_{i-\frac{1}{2}})+\mid c(x_{i-\frac{1}{2}})\mid)[u_h]_{i-\frac{1}{2}}(v_h)^+_{i-\frac{1}{2}}=0,$$
$$\int_{C_i}q_h(x,t)w_h(x)dx+\int_{C_i}\partial_x u_h(x,t)w_h(x)+[u]_{i+\frac{1}{2}}w(x^-_{i+\frac{1}{2}})=0 \quad \forall(x,t)\in\Omega\times[0,T]. \quad (4.67)$$

Using the DDG-scheme of diffusion and transport equation (4.58),(4.6), to get the matrix form for the DDG-schema (4.65). The RKDDG-schema for the equation (4.65) is given by:

$$\begin{cases} \dfrac{u^{n+1,i}-u^{n,i}}{\tau}+A^iq^{n,i}+B^iq^{n,i+1}+A^i_1u^{n,i}+B^i_1u^{n,i+1}=0, \\ q^{n,i}+D_1u^{n,i}+D_2u^{n,i-1}=0 \end{cases} \quad (4.68)$$

Where,

$$A^i=M^1(S^i_2-a(x_{i-\frac{1}{2}})P), \quad B^i=a(x_{i-\frac{1}{2}})M^1Q$$

$$A^i_1=M^1(S^i_1-b(x_{i+\frac{1}{2}})F+cM), \quad B^i_1=b(x_{i+\frac{1}{2}})M^1E$$

$$D_1=-M^1(S-F), \quad D_2=-M^1E$$

The matrix M, S_1, S_2, P, Q, E, F are given as follow:

$$M_{qk}=\int_{x_{i-\frac{1}{2}}}^{x_{i+\frac{1}{2}}}\varphi^i_k(x)\varphi^i_q(x)dx, \quad S_{qk}=\int_{x_{i-\frac{1}{2}}}^{x_{i+\frac{1}{2}}}\partial_x\varphi^i_k(x)\varphi^i_q(x)dx$$

$$(S_1)_{qk}=\int_{x_{i-\frac{1}{2}}}^{x_{i+\frac{1}{2}}}b(x)\partial_x\varphi^i_k(x)\varphi^i_q(x)dx, \quad (S_2)_{qk}=\int_{x_{i-\frac{1}{2}}}^{x_{i+\frac{1}{2}}}a(x)\partial_x\varphi^i_k(x)\varphi^i_q(x)dx$$

$$P=\varphi^i_k(x^+_{i-\frac{1}{2}})\varphi^i_q(x^+_{i-\frac{1}{2}}), \quad Q=\varphi^{i-1}_k(x^-_{i-\frac{1}{2}})\varphi^i_q(x^+_{i-\frac{1}{2}})$$

$$E=\varphi^i_k(x^+_{i+\frac{1}{2}})\varphi^i_q(x^-_{i+\frac{1}{2}}), \quad F=\varphi^i_k(x^-_{i+\frac{1}{2}})\varphi^i_q(x^-_{i+\frac{1}{2}})$$

We use $RK2$ and $RK3$ for the equation (4.68).

4.2.3 Computational results

Example 1: We provide preliminary numerical results for the schemes discussed in this section. We solve the Diffusion equation with a periodic boundary condition:

$$\begin{cases} u_t(x,t)-au_{xx}(x,t)=0, & \forall(x,t)\in[0,1]\times[0,1], \\ u_0(x)=sin(2\pi x), & \forall x\in[0,1]. \end{cases} \quad (4.69)$$

The exact solution is $u(x,t) = \exp(-a(2\pi)^2 t)\sin(2\pi x)$, and $a = 1.0$ Results for the L^∞ error and we use $RK2$, further corresponding orders are given in $Table3$. In Table 4, we check time accuracy for fixed h, we study the orders of the DG-RKp schemes when $p \in 1, 2, 3$.

	$K = 1$		$K = 2$		$K = 3$	
Mesh I	error	order	error	order	error	order
20	0.2918E-03	-	0.1248E-03	-	0.6897E-05	-
40	0.7523E-04	1.78	0.1554E-04	3.00	0.2470E-06	3.81
80	0.1984E-04	1.79	0.1940E-05	3.00	0.1743E-07	3.82
160	0.4845E-05	2.03	0.2448E-06	2.99	0.1098E-08	3.98

Table 4.6: L^∞ errors, Diffusion test with $u_0(x) = sin(2\pi x)$ and T=0.05, $\tau = 8.310^{-3}$

	Euler		RK2		RK3	
Mesh N	error	order	error	order	error	order
80	0.3682E-09	-	0.1365E-09	-	0.1263E-08	-
160	0.1780E-09	1.04	0.3522E-10	1.95	0.2109E-09	2.58
320	0.8293E-10	1.10	0.7384E-11	2.25	0.3050E-10	2.78
640	0.3939E-10	1.07	0.1569E-11	2.23	0.4903E-11	2.63

Table 4.7: Example 1(1D Diffusion)with fixed spatial order $K = 2$ and mesh $m = 100$, to check time accuracy

We test this example again, but with another parameter, we take for $T = 1.0$ and we use $RK3$, and the parameter N is dependant of the $CFL = 0.012$, the results given as bellow.

	Er L_2		L_{inf}	
Mesh I	error	order	error	order
10	1.01E-04	-	1.37E-04	-
20	1.26E-05	3.0	1.77E-05	2.95
40	1.58E-06	2.89	2.23E-06	2.28
80	1.98E-07	2.99	2.80E-07	2.99
160	1.47E-08	3.00	3.50E-08	2.99

Table 4.8: L^2,L^∞ errors, for Diffusion test with $K = 2$

Example 2:(1D convection diffusion). Now we consider a diffusion equation:

$$u_t(x,t) - \frac{1}{2}\sigma u_{xx}(x,t) + bu_x(x,t) = 0, \qquad \forall (x,t) \in [0,1] \times [0,1], \tag{4.70}$$

with initial data :

$$u_0(x) = cos(2\pi x) + \frac{1}{2}cos(4\pi x), \qquad \forall x \in [0,1].$$

Together with periodic boundary conditions on $(0,1), \sigma = 0.1$, $T = 1$, and b=0.3. The exact solution is given by

$$u(x,t) = \sum_{k=1,2} c_k exp(-2\sigma^2 k^2 \pi^2 t) cos(2k\pi(x - bt)),$$

with $c_1 = 1$ and $c_2 = \frac{1}{2}$.

	$K=2$		$K=3$	
Mesh I	error	order	error	order
10	1.403E-03	-	9.22E-04	-
20	1.809E-04	2.96	5.60E-06	-
40	2.270E-05	2.99	3.57E-07	4.00
80	2.842E-06	2.99	2.23E-08	4.00
160	3.550E-07	3.00	1.39E-09	4.00

Table 4.9: L_2 errors

Numerical example for European Option. For the numerical example we will consider $T = 0.5$, $r = 0.1$ and $\sigma = 0.2$, and $K = 100$.and $Smin = 60.653$; $Smax = 271.82$. For the Runge-Kutta test, is given in the table bellow :

	$K=1$		$K=2$		$K=3$	
Mesh I	error	order	error	order	error	order
10	0.1908E-03	-	2.35E-03	-	1.31E-04	-
20	0.5862E-03	1.70	3.09E-04	2.92	9.27E-06	3.82
40	0.1571E-03	1.97	3.90E-05	2.98	5.99E-07	3.95
80	0.3874E-04	2.02	4.87E-06	3.00	3.77E-08	3.98
160	0.9487E-05	2.03	6.09E-07	3.00	2.36E-09	4.00

Table 4.10: example 1(1D BS-Eurp)L^2 errors, for BSE test,CFL= 0.3 to check time accuracy, RK3

	Euler-Explicit		RK=2		RK=3	
Mesh N	error	order	error	order	error	order
80	0.3682E-09	-	0.1365E-09	-	0.1563E-08	-
160	0.1780E-09	1.04	0.3522E-10	1.95	0.2109E-09	2.88
320	0.8293E-10	1.10	0.8784E-11	2.00	0.2750E-10	2.97
640	0.3939E-10	1.07	0.2169E-11	2.23	0.3423E-11	3.00

Table 4.11: (B&S equation)with fixed spatial order k=2,

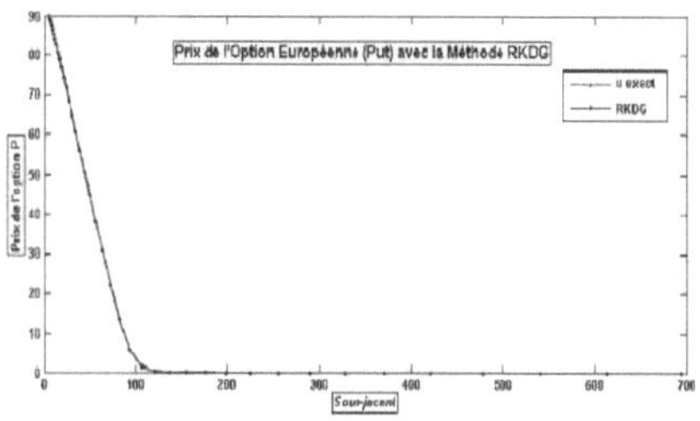

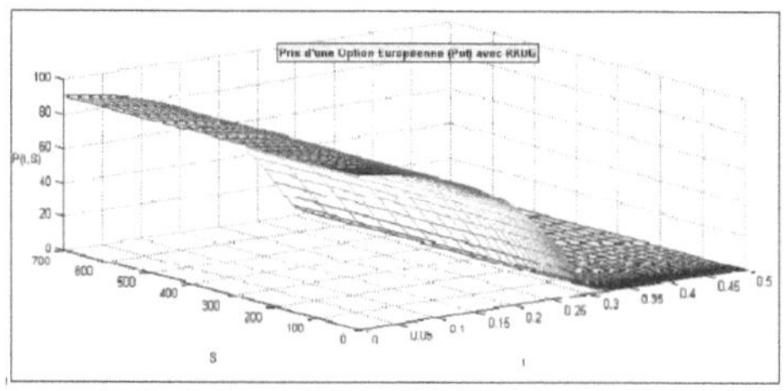

4.3 Conclusion

In this chapter, we have developed the extension of a discontinuous Galerkin method for solving diffusion-convection equation, approximating directly the solution variable rather than its derivatives as in the classical discontinuous Galerkin method. We have applied this new method to calculate the European option price by solving the Black-Scholes PDE.
The numerical experiments demonstrate that the method is stable and provides the optimal $(k+1)th$ order of accuracy for smooth solutions when using piecewise kth degree polynomials.
This method will be used for pricing the American put option :

$$\min(v_t - \frac{\sigma^2}{2}s^2 v_{ss} - rsv_s + rv,\ v - \varphi(s)) = 0, \quad t \in (0,T),\ s \in \mathcal{D} \tag{4.71a}$$
$$v(0,x) = \varphi(x) := \max(K - s, 0), \quad s \in \mathcal{D} \tag{4.71b}$$

with $\mathcal{D} = (S_{min}, S_{max})$, where $0 < S_{min} < K < S_{max}$ and S_{max} is large enough. Usually, working with continuous elements leads to a non trivial non linear equation to solve at each time step (even for the "explicit euler" scheme), of the form $\min(Mx - b, x - g) = 0$ (componentwise). The aim is to use the direct discontinuous Galerkin "DDG"scheme for the obstacle problem based on the work of [24](to our best knowledge), and for simplify significantly the implementation by taking the maximum at the Gaussian points. This new work is being finalized.

Chapter 5

Tikhonov-Type Regularization for Calibrating Stochastic Volatility in a Jump-Diffusion

Collaborated with : **R. Aboulaich**[1], **A. Jraifi**[2], **A. Darouichi**[3]

In this first part, we present a calibration problem that identifies the parameters of a stochastic-volatility jump-diffusion model using Tikhonov regularization, based on option prices—both model-computed and market-quoted. In general, closed-form pricing formulas are not available for these models. We formulate the calibration as an inverse problem for partial differential equations (PDEs), relying on a Dupire-type equation. We prove the existence of an optimal solution and also determine the analytic solution of the said equation. We emphasize that our approach applies to general stochastic-volatility jump-diffusion models, including local-stochastic-volatility jump-diffusions.

Many calibration models have been developed in recent years, for example volatility models (local volatility, stochastic volatility), but also models that account for jump risk (Lévy processes). In this work, we consider underlying assets whose price is modeled in multiple dimensions by a stochastic-volatility jump-diffusion, and we address the following question: is the model sufficiently regular to accurately reproduce prices across the whole financial asset universe, so that the model-determined option price is as close as possible to the market-quoted price?

Let $(\Omega, \mathcal{F}, (\mathcal{F}_t)_{t\geq 0}, \mathcal{P})$ be a filtered probability space, where $\mathcal{F}_t$ is the filtration of information available at time t and $\mathcal{P}$ is the real-world probability measure.

[1]EMI,Mohammed V University, Rabat
[2]ENSAS, Safi
[3]FST, Marrakech

We consider a financial asset whose price process $\{X_t, t \geq 0\}$ is given by the following stochastic differential equation with jumps (see [105]):

$$dX_t = F(t, X_t, Y_t)\, dt + G(t, X_t, Y_t)\, dW_t + \int_{\mathbb{R}^n} \gamma(t, X(t^-), z)\, \widetilde{N}^X(dt, dz), \qquad X_0 = x_0 \in \mathbb{R}^n \tag{5.1}$$

- $F : [0, T] \times \mathbb{R}^n \times \mathbb{R}^d \longrightarrow \mathbb{R}^n$ is the drift (transport) coefficient.
- $G : [0, T] \times \mathbb{R}^n \times \mathbb{R}^d \longrightarrow M_{n \times m}(\mathbb{R})$ is the diffusion (volatility) coefficient.
- $\gamma : [0, T] \times \mathbb{R}^n \times \mathbb{R}^n \longrightarrow M_{n \times l_X}(\mathbb{R})$.
- $\widetilde{N}^X(dt, dz) = (\widetilde{N}_1(dt, dz), \ldots, \widetilde{N}_{l_X}(dt, dz))$ are independent compensated Poisson random measures.
- W_t is a standard m-dimensional Brownian motion.

The process Y_t modeling stochastic volatility is given by the SDE:

$$dY_t = a(t, X_t, Y_t)\, dt + b(t, X_t, Y_t)\, dW_t^{'}, \qquad Y_0 = y_0 \in \mathbb{R}^d. \tag{5.2}$$

with:

- $W_t^{'}$ is a standard m-dimensional Brownian motion, possibly correlated with W_t. We assume this correlation is constant and denote it by $\rho \in [-1, 1]$, so that $d\langle W, W' \rangle_t = \rho\, dt$. If we define Z_t by $W_t^{'} = \rho W_t + \sqrt{1 - \rho^2}\, Z_t$, then W_t and Z_t are independent Brownian motions.
- $a : [0, T] \times \mathbb{R}^n \times \mathbb{R}^d \longrightarrow \mathbb{R}^d$ is the drift operator of the process Y.
- $b : [0, T] \times \mathbb{R}^n \times \mathbb{R}^d \longrightarrow M_{d \times m}(\mathbb{R})$ is the diffusion operator.

We propose an analytical approach based on integro-differential equations and market data $U^*(K, T)$ (European option prices observed at time t^*, corresponding to strikes K and maturities T) to calibrate the volatility dynamics $Y(t)$. Assuming the function $b(\cdot, \cdot)$ is known, estimating the volatility Y amounts to identifying the drift term $a(\cdot, \cdot)$ from the market option prices $U^*(K, T)$.

More precisely, we consider the following stochastic integro-differential system:

$$dX_t = F(t, X_t, Y_t)\, dt + \sigma_t X_t\, dW_t + \int_{\mathbb{R}^n} \gamma(t, X(t^-), z)\, \widetilde{N}^X(dt, dz), \tag{5.3}$$

$$\sigma_t = f(Y_t), \tag{5.4}$$

$$dY_t = a(t, Y_t)\, dt + b(t, Y_t)\, dW_t^{'}, \tag{5.5}$$

$$X_0 = x_0 \in \mathbb{R}^n, \tag{5.6}$$

$$Y_0 = y_0 \in \mathbb{R}^d. \tag{5.7}$$

Calibration in this setting consists in identifying the term $a(t,y)$ and $\omega(t,x,y;K,T)$ solving $(6.18)_1$–$(6.18)_2$ such that

$$\omega(t^*,x^*,y^*;K,T) = U^*(K,T), \quad \text{for all } K,T,$$

where ω solves the following PDE problem:

$$\begin{cases} \text{Find } \omega \text{ such that:} & & \\ \dfrac{\partial w}{\partial t}(t,x,y) + \mathcal{L}w(t,x,y) = 0 & \forall t \in [0,T],\ x \in \mathbb{R}^n,\ y \in \mathbb{R}^d, \\ w(T,x,y) = h(x) & x \in \mathbb{R}^n,\ y \in \mathbb{R}^d. \end{cases} \tag{5.8}$$

where $\mathcal{L}$ is the infinitesimal generator associated with the Markov diffusion (X_t, Y_t), namely:

$$\begin{aligned} \mathcal{L}(\omega) &= \frac{1}{2}\sum_{i,j=1}^{n}(GG^T)_{ij}\frac{\partial^2\omega}{\partial x_i \partial x_j} + \sum_{i=1}^{n} F_i \frac{\partial \omega}{\partial x_i} - r\,\omega \\ &+ \int_{\mathbb{R}^n}\sum_{k=1}^{l_X}\left\{\omega(x+\gamma^{(k)}(t,x,z)) - \omega(x) - \gamma^{(k)}(t,x,z)\cdot\nabla\omega\right\}\nu_k^X(dz_k) \\ &+ \frac{1}{2}\sum_{i,j=1}^{d}(bb^T)_{ij}\frac{\partial^2\omega}{\partial y_i \partial y_j} + \sum_{i=1}^{d} a_i\frac{\partial\omega}{\partial y_i} + \rho\sum_{i,j=1}^{n,d}(Gb^T)_{ij}\frac{\partial^2\omega}{\partial x_i \partial y_j} + \sum_{i=1}^{d} R(t,x,y)\frac{\partial \omega}{\partial y_i} \end{aligned} \tag{5.9}$$

- $R(t,X,Y) = \rho\,\dfrac{F(t,X,Y)-r}{G(t,X,Y)} + \sqrt{1-\rho^2}\,\theta^*(t,x,y)$, where $\theta^*(t,x,y)$ is an arbitrary function; see, for example, ([11], [?]).

For details on existence and uniqueness of the solution, see ([?], [?]).

The remainder of this work is developed and presented in ([?]) as follows: after introducing the calibration problem, we determine the fundamental solution of the standard inverse problem, following the Dupire approach. We then study this inverse problem via Tikhonov regularization. This approach enables the stable reconstruction of the parameters of a stochastic-volatility jump-diffusion model from the observation of a finite set of options. We prove the existence of a minimizer of the cost functional. Moreover, we obtain the analytic solution of the said equation and, finally, we establish the uniqueness and stability of the minimizer.

CHAPTER 6

Source-Term Identification Problem in an Unconfined Aquifer

Collaborated with : **S. Slimani**[1], **K. Najib**[2], **A. Zinz** [3]

Contents

6.1 Introduction

In this chapter, we address the identification of the source term for problem (??), using boundary data on h_1. Here, h_1 denotes the depth of the aquifer?s free interface. We apply a variable-separation technique used in [91].

[1]EMI,Mohammed V University, Rabat

[2]Mines Rabat

[3]Ecole Centrale Lyon

Why identify the source term?

Identifying the source term in saltwater intrusion problems plays an important role. This process helps water-resource managers develop better strategies to protect people and their water supplies.
Mathematically, the source-identification problem is an inverse problem. The direct problem, for an unconfined aquifer consists in finding the unknowns h, the depth of the freshwater/saltwater interface, and h_1, the depth of the free interface, given the source terms, model parameters such as conductivity and porosity, and boundary and initial conditions. In our case the source term is unknown and must be determined, is called the inverse of the direct problem. According to Hadamard, a problem is well-posed if

1. a solution exists,
2. the solution is unique,
3. the solution depends continuously on the data.

A problem failing any of these three conditions is called ill-posed.
Inverse problems are generally ill-posed, and the problem we treat here is no exception. Two questions then arise:

1. Can the source term be uniquely identified from a few measurements of the depth h_1?
2. The available data/measurements of h_1 are noisy, which leads to instability of the solution; how can we address this?

Formulation of the source-identification problem

For source identification we apply the separation-of-variables technique used in [91]. The method requires a linear problem, whereas model (2.27) is non-linear. We therefore adopt a linearization procedure and prove a convergence result.

In the linearized model, as in [91], we assume the source term splits as $F(x)H(t)$, where $F(x)$ represents the well location and $H(t)$ the pumping rate over time. Thus, at each step of the linearization algorithm, we focus on identifying $F(x)$.

Unlike [91], we have non-homogeneous boundary data and initial conditions for the solution h_1. We therefore write the linearized problem as the sum of two subproblems: P_1 with homogeneous boundary and initial conditions, and P_2 with non-homogeneous ones. At this point, we can reconstruct $F(x)$

uniquely from data $h_1 = u_1 + u_2$, where u_i denotes the boundary data for problem P_i, $i = 1, 2$. The method consists of solving a Sturm-Liouville problem; its solutions form a basis in which $F(x)$ is expanded as a series. Hence, identifying $F(x)$ amounts to identifying the series coefficients.

Noise in the h_1 data causes instability. We use Tikhonov regularization to mitigate this. The L-curve method is employed to choose a suitable regularization parameter β. The problem reduces to minimizing the following cost functional:

$$\mathfrak{J}_\beta(F) = \tfrac{1}{2}\,\|h_1(L,t;\ \mathbf{F}\ ) - g_\varepsilon(t)\|^2_{L^2(0,T)} + \frac{\beta}{2}\,\|\ \mathbf{F}\ \|_2^2,$$

where $g_\varepsilon(t)$ denotes noisy data.
To achieve the chapter?s goal, we proceed as follows. In Section 6.2, using a fixed-point method, we prove a convergence result for the direct problem. Next, in Section 6.3, we set the mathematical framework in which the inverse problem is solved and present the source-identification procedure. Finally, in Section 6.4, we provide numerical results for both the direct and inverse problems.

6.2 Linearization and convergence result for the direct problem

To apply the separation-of-variables technique of [91] for reconstructing an x-dependent source, we transform problem (2.27) into a sequence of linear equations via a fixed-point algorithm.

The linearization is as follows: given an initial value (h_1^0, h^0) such that

$$(h_1^0 - h_{1,D}, h^0 - h_D) \in W(0,T) \times W(0,T),$$

for all $n \geq 0$ we seek (h_1^{n+1}, h^{n+1}) with $(h_1^{n+1} - h_{1,D}, h^{n+1} - h_D) \in [W(0,T)]^2$, a weak solution of

$$\begin{cases} \phi\partial_t h_1^{n+1} - \operatorname{div}\big(K(T_f(h^n - h_1^n) + T_s(h^n))\nabla h_1^{n+1}\big) - \operatorname{div}\big(KT_s(h^n)\nabla h^n\big) - \operatorname{div}(\delta\phi\nabla h_1^{n+1}) \\ \qquad\qquad = -\,Q_sT_s(h^n) - Q_fT_f(h^n - h_1^n), \\ \phi\partial_t h^{n+1} - \operatorname{div}\big(KT_s(h^n)\nabla h^{n+1}\big) - \operatorname{div}\big(KT_s(h^n)\nabla h_1^n\big) - \operatorname{div}(\delta\phi\nabla h^{n+1}) \\ \qquad\qquad = -\,Q_sT_s(h^n). \end{cases} \tag{6.1}$$

Theorem 4.

For each $n \geq 0$, under the assumptions:

- weak spatial heterogeneity of the hydraulic conductivity tensor,

$$\frac{K^{+^2}}{K^-} \leq \frac{\delta\phi}{h_2}, \tag{6.2}$$

- $(h_1^n - h_{1,D}, h^n - h_D) \in W(0,T) \times W(0,T)$,
- $\|\nabla(h_1^n - h_{1,D})\|^2_{L^2(0,T,H^1)} \leq \sqrt{\frac{M_0}{\delta\phi}}, \quad \|\nabla(h^n - h_D)\|^2_{L^2(0,T,H^1)} \leq \sqrt{\frac{M_0}{\delta\phi}}$,
- $0 \leq h_1^n(t,x) \leq h^n(t,x) \leq h_2$ a.e. $x \in \Omega$ for all $t \in [0,T]$,

problem (6.1) admits a weak solution (h_1^{n+1}, h^{n+1}) with $(h_1^{n+1} - h_{1,D}, h^{n+1} - h_D) \in [W(0,T)]^2$.
Moreover, as $n \to \infty$, h^{n+1} (resp. h_1^{n+1}) converges almost everywhere to h (resp. h_1) in $[0,T] \times \Omega$, where (h, h^{n+1}) is the unique solution of (**??**).

Remark 6.2.1. *The proof presented here follows [39] for confined aquifers, and $M_0 > 0$ depends on the problem data.*
Assumption (6.2) is reasonable because the aquifer?s hydraulic conductivity is small.

We use a fixed-point method to prove thm 4. The proof proceeds in several steps:

1. show that h_1^{n+1}, h^{n+1} are uniformly bounded in

$$L^2(0,T,H^1(\Omega)) \cap L^\infty(0,T,H);$$

2. prove that $(\partial_t(h_1^{n+1} - h_{1,D}))_n$ and $(\partial_t(h^{n+1} - h_D))_n$ are bounded in $L^2(0,T,V')$;
3. determine the constant M_0;
4. by Aubin?s lemma, extract a subsequence converging strongly in $L^2(\Omega_T)$ and weakly in $L^2(0,T,H^1(\Omega)) \cap H^1(0,T,V')$.

Proof.
Step 1.
(The weak formulations, estimates, Gronwall inequality, and resulting bounds

are kept identical to your original text; only explanatory prose has been translated. Equations (??)?(??) unchanged.)

Step 2.
(We show $(\partial_t(h_1^{n+1} - h_{1,D}))_n$ is bounded in $L^2(0,T,V')$ via (??). Estimates (??) follow verbatim.)

Step 3.
(Using test functions $h_1^{n+1} - h_{1,D}$ and $h^{n+1} - h_D$, summing (??)?(??), and estimating $E_1?E_{10}$ yields the recursive bound, from which M_0 is defined. Details as in the original text.)

Step 4.
(Compactness via Aubin?Lions and uniqueness give convergence to the unique solution (h, h_1) of (??); see (??)?(??).) □

6.3 Identification of the source term

6.3.1 Notation and assumptions

The technique in [91] is studied in one dimension. Hence we take $\Omega = [0, L]$. Let

- $T_1(x) = K(T_f(h^n - h_1^n) + T_s(h^n) + \delta\phi)$,
- $S^n(x) = \partial_x\big(KT_s(h^n)\partial_x h^n\big)$,
- $F^n(x)H(t) = Q_f T_f(h^n - h_1^n)$.

At each fixed-point step n, we identify the unknown source $F^n(x)$ for the linear problem

$$\begin{cases} \phi\partial_t h_1^{n+1} - \partial_x\big(T_1(x)\partial_x h_1^{n+1}\big) = S^n(x) + F^n(x)H(t), \\ h_1^{n+1}(0,t) = h_{1,D}(0,t),\ \partial_x h_1^{n+1}(L,t) = 0 \quad \forall t \in [0,T], \\ h_1^{n+1}(x,0) = h_{1,0}(x) \quad \forall x \in \Omega, \end{cases} \tag{6.3}$$

from the boundary data $g(t) = h_{1,D}(L,t)$.

Remark 6.3.1. *$T_1(x)$ is positive by the maximum principle, since*

$$0 \le h_1^n \le h^n \le h_2.$$

We write

$$h_1^{n+1}(x,t) = U^{n+1} + V^{n+1},$$

where U^{n+1} solves

$$\begin{cases} \phi\partial_t U^{n+1} - \partial_x\big(T_1(x)\partial_x U^{n+1}\big) = S^n(x), \\ U^{n+1}(0,t) = h_{1,D}(0,t),\ \partial_x U^{n+1}(L,t) = 0, \quad U^{n+1}(x,0) = h_{1,0}(x), \end{cases} \tag{6.4}$$

and V^{n+1} solves

$$\begin{cases} \phi\partial_t V^{n+1} - \partial_x\big(T_1(x)\partial_x V^{n+1}\big) = F^n(x)H(t), \\ V^{n+1}(0,t) = \partial_x V^{n+1}(L,t) = 0,\ V^{n+1}(x,0) = 0. \end{cases} \tag{6.5}$$

Thus $g(t) = g_u(t) + g_v(t)$ with $g_u(t) = U^{n+1}(L,t)$ and $g_v(t) = V^{n+1}(L,t)$. Using separation of variables in (6.5),

$$V^{n+1}(x,t) = \sum_{k\geq 1} A_k(t)B_k(x), \tag{6.6}$$

where $B_k(x)$, $k \geq 1$, solve the regular Sturm?Liouville problem

$$\begin{cases} -\partial_x\big(T_1(x)\partial_x B(x)\big) = \lambda B(x), \\ B(0) = \partial_x B(L) = 0. \end{cases} \tag{6.7}$$

The eigenvalues λ_k satisfy $0 < \lambda_1 < \lambda_2 < \cdots \to +\infty$. The functions B_k form a complete orthogonal family in $L^2(\Omega)$, normalized to an orthonormal basis χ_k, $k \geq 1$. The source is expanded as

$$F^n(x) = \sum_{k=1}^{+\infty} f_k \chi_k(x), \tag{6.8}$$

with $f_k = \langle F^n(x), \chi_k\rangle_{L^2(0,L)}$. Substituting (6.6) and (6.8) into (6.5) yields

$$\begin{cases} \phi A_k'(t) + \lambda_k A_k(t) = \langle F^n, \chi_k\rangle_{L^2(0,L)}\, H(t), \\ A_k(0) = 0, \end{cases} \tag{6.9}$$

so

$$A_k(t) = \frac{1}{\phi} f_k\, e^{-\frac{\lambda_k}{\phi}t}\left(\int_0^t H(s)\, e^{\frac{\lambda_k}{\phi}s} ds\right). \tag{6.10}$$

Hence

$$V^{n+1}(x,t) = \sum_{k\geq 1} \frac{\chi_k(x)}{\phi}\, f_k\, e^{-\frac{\lambda_k}{\phi}t}\left(\int_0^t H(s)\, e^{\frac{\lambda_k}{\phi}s} ds\right). \tag{6.11}$$

6.3.2 Uniqueness of source identification

Lemma 2.

If $H(t) \in C^1[0,T]$ is known with $H(0) \neq 0$, then $F^n(x) \in L^2(\Omega)$ can be uniquely identified from the boundary data $g(t)$.

Proof.
The proof follows the argument in [91]. □

6.3.3 Reconstruction of the source term

Reconstructing the unknown $F^n(x)$ amounts to determining the coefficients $f_k = \langle F^n, \chi_k \rangle_{L^2(0,L)}$ in

$$V^{n+1}(L,t) = \sum_{k\geq 1} f_k \, \Psi_k(t), \tag{6.12}$$

where $\Psi_k(t) = \dfrac{\chi_k(L)}{\phi} \int_0^t H(s)\, e^{-\frac{\lambda_k}{\phi}(t-s)} ds$.
We truncate (6.12) and approximate

$$V^{n+1}(L,t) \simeq \sum_{k=1}^{N} f_k \, \Psi_k(t). \tag{6.13}$$

(The series (6.12) satisfies the Cauchy criterion.)
Thus the inverse problem reduces to a finite-dimensional one: identify the N coefficients f_k in (6.13) from the boundary data $g(t)$.
Experimental measurements may be noisy, $g^{\varepsilon}(t)$, with $\|g - g^{\varepsilon}\| \leq \varepsilon$. We therefore minimize the mismatch between $h_1^{n+1}(L,t)$ and the noisy data $g^{\varepsilon}(t)$ using the cost

$$\mathfrak{J}(F^n) = \frac{1}{2} \left\| \left(g_u(t) + V^{n+1}(L,t)\right) - g^{\varepsilon}(t) \right\|^2_{L^2(0,T)}.$$

Since reconstruction from noisy data is unstable, we adopt Tikhonov regularization and minimize

$$\begin{aligned} \mathfrak{J}_\beta(F^n) =& \frac{1}{2} \left\| \left(g_u(t) + V^{n+1}(L,t)\right) - g^{\varepsilon}(t) \right\|^2_{L^2(0,T)} + \frac{\beta}{2} \|F^n\|_2^2 \\ =& \frac{1}{2} \int_0^T \left(g_u(t) + \sum_{k=1}^{N} \mathbf{f}_k \, \Psi_k^{n+1}(t) - g^{\varepsilon}(t) \right)^2 dt + \frac{\beta}{2} \sum_{k=1}^{N} \mathbf{f}_k^2, \end{aligned} \tag{6.14}$$

with $\beta > 0$ the regularization parameter.

Remark 6.3.2. *Tikhonov method.*

To find the minimizer, differentiate (6.14) (Lagrangian equation):

$$\frac{\partial \mathcal{J}_\beta}{\partial f_j} = \int_0^T \sum_{k=1}^N f_k\, \Psi_k(t)\, \Psi_j(t)\, dt + \int_0^T g_u(t)\, \Psi_j^{n+1}(t)\, dt - \int_0^T g^\varepsilon(t)\, \Psi_j^{n+1}(t)\, dt + \beta f_j$$
$$=0, \qquad \forall 1 \leq j \leq N. \tag{6.15}$$

In matrix form,

$$(A + \beta I_N)\mathbf{F}^n = b, \tag{6.16}$$

where

- $A_{jk} = \int_0^T \Psi_k(t)\, \Psi_j(t)\, dt, \quad j = 1, \ldots, N, \ k = 1, \ldots, N,$
- $b_j = \int_0^T \left(g^\varepsilon(t) - g_u(t)\right) \Psi_j(t)\, dt, \quad j = 1, \ldots, N.$

6.4 Numerical simulations of the source-reconstruction algorithm

This section presents numerical results for a 1D source-reconstruction problem using synthetic data, since no real data are available. We first solve the direct problem to obtain the interface evolution. We take the free-interface value at $x = 60\,\mathrm{m}$ as data.
We report numerical results for the direct problem in 6.4.1, followed by results for the source-reconstruction problem in 6.4.2. The computations were performed in MATLAB.

6.4.1 Interface-displacement problem

The aquifer is approximated in 1D with $\Omega = [0, 60\,\mathrm{m}]$.
The discretization set is $\{x_i \in [0, 60] : x_i = i\,h,\ 0 \leq i \leq N_p\}$. Problem inputs:

- hydraulic conductivity $K = 39.024\,\mathrm{m/d}$,
- porosity $\phi = 0.3$,
- pumping near $x_0 = 35$ with constant rate $0.6\,\mathrm{m/d}$.

Figure 6.1 will be shown last, just before starting the inverse problem. Two additional 3D figures will be added for pumping and injection cases. Result: under pumping, the aquifer level decreases

while the saltwater volume increases; under injection, the opposite occurs. Figure 6.1 is provisional.

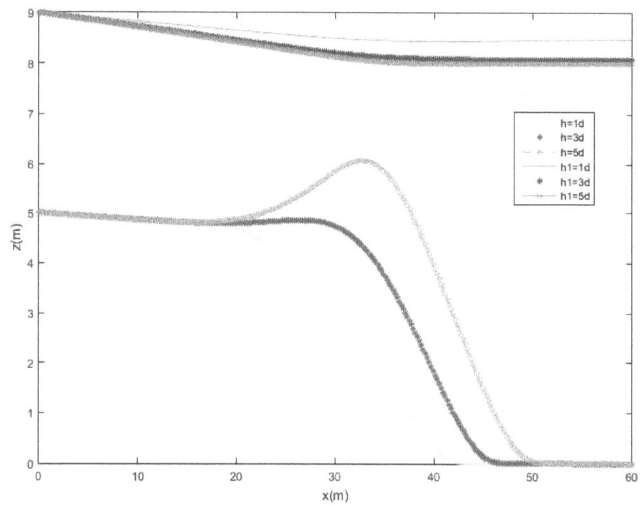

Figure 6.1: Evolution of the interface (h) and upper surface (h_1) at times $T = 1,\ 3,\ 5$ days with $Q_f = -0.6 \times \exp\{-0.01\,(x-35)^2\}$.

Thus we obtain the approximate solution of (??) at the boundary $h_{1,D}(60,t)$ for $t \in [0,T]$, $T = 1$ day and $T = 5$ days.
In Figure 6.1, a saltwater cone forms in the pumping zone while the aquifer level drops.

6.4.2 Source-reconstruction problem

Source reconstruction requires eigenvalues $\lambda > 0$ such that an eigenfunction vector $\mathbf{B} = \left(B(x_i),\, i = \overline{0, N_p}\right)$ solving (6.7) exists.
Since the eigenspace is 1D, to ensure uniqueness of eigenfunctions we impose the normalization

$$\int_0^L |B(s)|^2\, ds = 1.$$

For numerical computation of eigenvalues and eigenfunctions, we rewrite (6.7) by introducing two auxiliary quantities. We treat λ as a function of x and add

$$\lambda'(x) = 0.$$

Define

$$Z(x) = \int_0^x |B(s)|^2\, ds,$$

giving another ODE and two additional boundary conditions:

$$Z'(x) = |B(x)|^2, \quad Z(0) = 0, \quad Z(L) = 1.$$

We thus obtain a boundary-value problem for three unknowns, $B(x)$, $\lambda(x)$, and $Z(x)$. To derive a first-order ODE system, apply the transform

$$(y_1(x), y_2(x)) = (B(x), B'(x))$$

to (6.7), and set

$$y_3(x) = \lambda(x), \qquad y_4(x) = Z(x).$$

This yields

$$\mathbf{y}'(x) = \begin{bmatrix} 0 & 1 & 0 & 0 \\ 0 & -\dfrac{T_1'(x)}{T_1(x)} & 0 & 0 \\ 0 & 0 & 0 & 0 \\ 0 & 0 & 0 & 0 \end{bmatrix} \mathbf{y}(x) + \begin{bmatrix} 0 \\ -\dfrac{1}{T_1(x)}\, y_1(x)\, y_3(x) \\ 0 \\ y_1^2(x) \end{bmatrix}, \tag{6.17}$$

$$\begin{bmatrix} 1 & 0 & 0 & 0 \\ 0 & 0 & 0 & 0 \\ 0 & 0 & 0 & 1 \\ 0 & 0 & 0 & 0 \end{bmatrix} \mathbf{y}(0) + \begin{bmatrix} 0 & 0 & 0 & 0 \\ 0 & T_1(L) & 0 & 0 \\ 0 & 0 & 0 & 0 \\ 0 & 0 & 0 & 1 \end{bmatrix} \mathbf{y}(L) = \begin{pmatrix} 0 \\ 0 \\ 0 \\ 1 \end{pmatrix}, \tag{6.18}$$

with $\mathbf{y} = (y_1, y_2, y_3, y_4)$.
To solve (6.17)?(6.18) numerically, we use `bvp5c` in MATLAB (see [?]), a finite-difference collocation method implementing 4-stage Lobatto IIIA. See [123] for details.
We compute the relative error

$$\varepsilon_F = \frac{\|F_{\text{ex}} - F_{\text{rec}}\|_{L^2([0,50])}}{\|F_{\text{ex}}\|_{L^2([0,50])}}.$$

To balance accuracy and cost, we use 26 eigenpairs in the reconstruction; Figure 6.2 shows that $\|F_{\text{rec},K+1} - F_{\text{rec},K}\|$ decreases rapidly for $K = 26$.

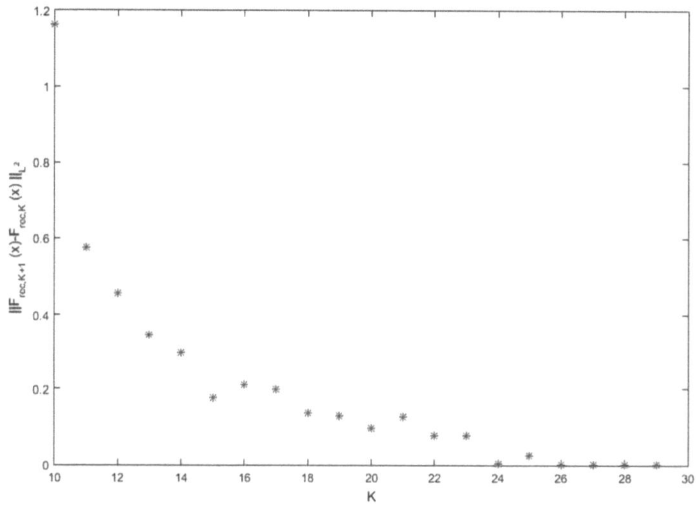

Figure 6.2: L^2-norm of the difference between space-dependent reconstructed sources using $K+1$ vs K eigenfunctions in the numerical implementation for $\phi = 0.3$.

We choose the optimal β via the L-curve method [74], plotting (log?log) the squared norm of the regularized solution $\|F\|^2_{L^2}$ against the squared residual $\|AF - b\|^2_{L^2}$ for a range of β; the ?corner?? gives the best trade-off. In our case, $\beta = 1 \cdot 10^{-13}$ for $\phi = 0.3$.
To illustrate sensitivity to β, we also test $\beta = 3 \cdot 10^{-13}$ and $\beta = 9 \cdot 10^{-13}$; the reconstruction degrades away from the optimal value.

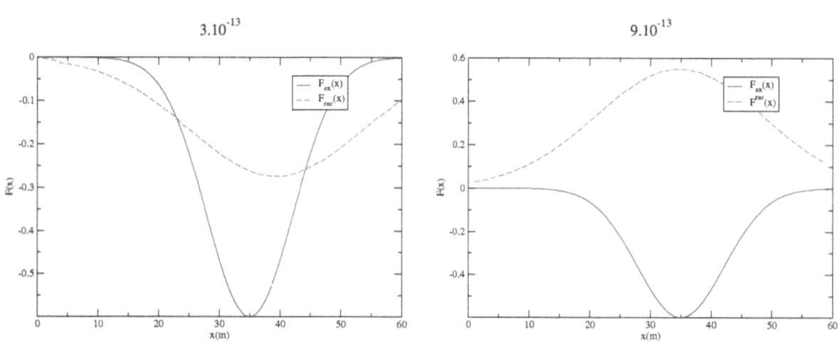

Figure 6.3: Source reconstruction at 1 day for β values different from the optimal one.

Table 6.1 reports errors for $T = 1$ and $T = 5$ days. The Tikhonov regularization parameter is thus crucial for the inverse problem.

	1 day			5 days		
β	$9 \cdot 10^{-13}$	$1 \cdot 10^{-13}$	$3 \cdot 10^{-13}$	$5 \cdot 10^{-13}$	$5 \cdot 10^{-14}$	$1 \cdot 10^{-14}$
ε_F	0.5445	0.0803	0.7971	0.4276	0.1465	2.0366

Table 6.1: Regularization parameters β and relative errors ε_F at 1 and 5 days.

Source reconstruction

Based on the previous results?$K = 26$ eigenfunctions and an L-curve?selected β?we reconstruct for a final time of 5 days. Figure 6.4 compares the reconstructed source F_{rec} and the exact F_{ex} at 1 and 5 days across the domain. The approximations agree well with the truth. We note that 6 iterations were needed after 1 day, and 29 iterations after 5 days.

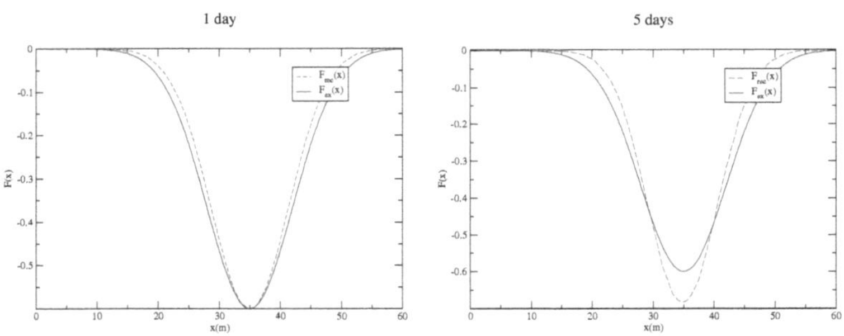

Figure 6.4: Reconstruction of the space-dependent source at 1 day and 5 days, $\phi = 0.3$.

Simulated data

We compute the free interface at $x = 60\,\text{m}$ using the reconstructed source F_{rec} in Figure 6.5. The left panel zooms on the 1-day period. The results agree well with the data. A quantitative comparison using the relative L^2 error between the two solutions yields 4.5×10^{-3}, validating the approach.

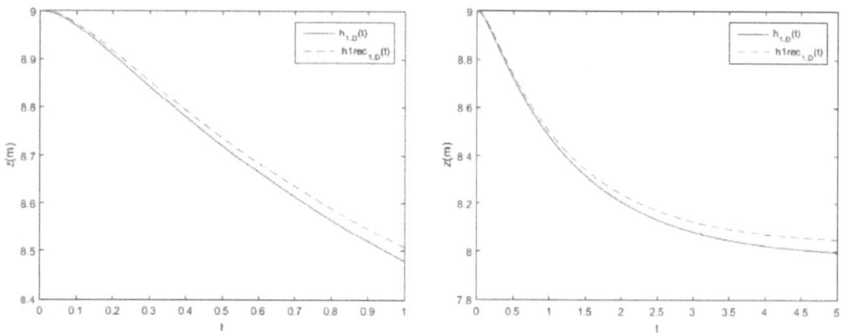

Figure 6.5: Free interface computed using the reconstructed source $h_{\text{rec},1,D}(60,t)$ versus synthetic data $h_{1,D}(60,t)$ over time: 1 day (left) and 5 days (right), $\phi = 0.3$.

Impact of physical parameters

Table 6.2 shows the relative reconstruction error for different porosities. For lower porosities, reconstruction improves; for highly permeable domains, accuracy may degrade. This may be due to model-change uncertainties, since Darcy?s law becomes less valid in those regimes. The regularization parameter is selected by the L-curve criterion in each case.

	$\phi = 0.3$	$\phi = 0.5$	$\phi = 0.8$
β	$1 \cdot 10^{-13}$	$3 \cdot 10^{-13}$	$2 \cdot 10^{-13}$
ε_F	0.0803	0.3059	0.5410
Iterations	6	3	4

Table 6.2: Relative error ε_F for different porosity values ϕ at 1 day.

6.5 Conclusion

We studied an inverse problem of reconstructing the source from measurements of the free interface at the domain boundary. The original problem was linearized so that the reconstruction method of [91] could be applied, and a convergence result was established. The inverse problem reduces to a 1D Sturm?Liouville problem. Because the problem is unstable, Tikhonov regularization was adopted. Numerically, we obtained good source reconstructions under suitable parameter conditions, using an appropriate number of eigenfunctions and an L-curve?based choice of the regularization parameter. Numerical convergence of the source term was demonstrated.

CHAPTER 7

Mixed formulation of an intrusion problem in a confined aquifer

Collaborated with : **M. Farhloul**[1], **S. Slimani**[2], **K. Najib**[3], **A. Zinz** [4]

Contents

7.1 Introduction

This chapter develops a mixed formulation tailored to the seawater-intrusion problem in a confined aquifer, adopting a sharp/diffuse interface viewpoint.

[1]Moncton University
[2]EMI,Mohammed V University, Rabat
[3]Mines Rabat
[4]Ecole Centrale Lyon

In the first part, we recast problem (??) from Chapter ?? into a mixed form by introducing a new variable σ such that

$$\begin{cases} M\,\partial_t u - \operatorname{div}\,\sigma & = F, \\ \sigma - R(u)\,\nabla u & = 0, \end{cases}$$

where $u = \begin{pmatrix} h \\ \Phi_f \end{pmatrix}$ and $R(u)$ is a symmetric positive definite matrix.

In the second part, we establish several properties of $R(u)$. Building on prior results, we prove that the stationary mixed problem admits a unique solution.

In the fourth part, we consider the discrete setting for the mixed problem. We linearize the system and prove a linear convergence result for the iterative solution. We then derive an a priori error estimate.

7.2 Mixed formulation of the saline intrusion problem

In this section, we rewrite system (??) in mixed form. Relying on results previously obtained for that system, we derive auxiliary properties that ensure existence and uniqueness for the mixed formulation.

Notation and assumptions

Let $\Omega \subset \mathbb{R}^2$ be a bounded open set representing the horizontal projection of the porous medium, with a boundary Γ of class $\mathcal{C}^1$. The time interval is $(0, T)$ and we set $\Omega_T = (0, T) \times \Omega$.
We work with model (??) under the assumptions and results stated in the modeling chapter. We assume the maximum principle holds and restrict to the case where K is scalar (isotropic). The problem reads

$$\begin{cases} \phi\,\partial_t h + \nabla\cdot\big(K(h_2 - h)\,\nabla\Phi_f\big) - \nabla\cdot\big(\alpha K(h_2 - h)\,\nabla h\big) - \phi\,\nabla\cdot(\delta\,\nabla h) & = -Q_s, \\ -\nabla\cdot\big(K\,h_2\,\nabla\Phi_f\big) + \nabla\cdot\big(\alpha K(h_2 - h)\,\nabla h\big) & = Q_f + Q_s, \end{cases} \tag{7.1}$$

where Q_i, $(i = f, s)$ denote external sources corresponding to pumping or recharge by fresh and saline water, respectively.

7.2.1 Mixed problem

Our aim is to write a mixed formulation of (7.1) and then prove well-posedness.

Multiplying the second equation of (7.1) by $\frac{1}{\alpha}$ gives

$$\begin{cases} \phi\,\partial_t h + \nabla\cdot\big(K(h_2-h)\,\nabla\Phi_f\big) - \nabla\cdot\big(\alpha K(h_2-h)\,\nabla h\big) - \phi\,\nabla\cdot(\delta\nabla h) = -Q_s, \\ -\nabla\cdot\big(\frac{K}{\alpha}\,h_2\,\nabla\Phi_f\big) + \nabla\cdot\big(K(h_2-h)\,\nabla h\big) = \frac{1}{\alpha}(Q_f+Q_s). \end{cases} \tag{7.2}$$

In matrix form, (7.2) becomes

$$\begin{cases} M\,\partial_t u - \operatorname{div}\sigma &= F \quad \text{in } \Omega_T, \\ \sigma - R(u)\,\nabla u &= 0 \quad \text{in } \Omega_T, \\ u &= u_D \quad \text{on } \Gamma, \end{cases} \tag{7.3}$$

with

$$M = \begin{bmatrix} \phi & 0 \\ 0 & 0 \end{bmatrix}, \quad u = \begin{pmatrix} h \\ \Phi_f \end{pmatrix}, \quad F = \begin{pmatrix} -Q_s \\ \frac{1}{\alpha}(Q_f+Q_s) \end{pmatrix},$$

$$R(u) = \begin{bmatrix} \delta\phi - \alpha K(h-h_2) & K(h-h_2) \\ K(h-h_2) & \frac{K}{\alpha}h_2 \end{bmatrix}, \quad \nabla u = \begin{bmatrix} \partial_x h & \partial_y h \\ \partial_x\Phi_f & \partial_y\Phi_f \end{bmatrix}, \quad \text{and}$$

$$u_D = \begin{pmatrix} h_D \\ \Phi_{f,D} \end{pmatrix}.$$

7.2.2 Mathematical tools

We now collect results used to prove existence and uniqueness for (7.3).

7.2.2.1 Properties of $R(u)$

Proposition 4.

Let $R(u) = \begin{bmatrix} a & b \\ b & c \end{bmatrix}$ with $a = \delta\phi - \alpha K(h - h_2)$, $b = K(h - h_2)$, and $c = \frac{K}{\alpha} h_2$.
Then:

1. $R(u)$ is symmetric positive definite.
2. The eigenvalues of $R(u)$ are bounded independently of u.

Proof.

1. Since $0 < \delta \leq h \leq h_2$, we have $a > 0$ and
$$\det(R(u)) = \frac{\delta\phi K}{\alpha} h_2 + K^2 h(h_2 - h) > 0,$$
so Sylvester?s criterion yields positive definiteness.
2. As $R(u)$ is SPD, its eigenvalues $0 < \lambda_1 < \lambda_2$ satisfy
$$K\delta \leq \lambda_1 < \lambda_2 < \delta\phi + K h_2\left(\tfrac{1}{\alpha} + 1\right),$$
hence they are uniformly bounded in u.

□

7.2.2.2 Properties of $R(u)^{-1}$

Let $S(u) := R(u)^{-1}$.

Proposition 5.

The matrix $S(u)$ is symmetric positive definite and:

1. Its eigenvalues are bounded independently of u.
2. S is L-Lipschitz in u:
$$\|S(u) - S(v)\|_{L^\infty} \leq L\|u - v\| \quad \forall u, v \in [L^2(\Omega)]^2,$$
with L depending only on the data.

Proof. Immediate from the bounds on the eigenvalues of $R(u)$ and a direct computation of $S(u)$; details as in the original text.

□

We next analyze the stationary case of (7.3). Without loss of generality, we take homogeneous Dirichlet data $u_D = 0$.

7.3 Stationary mixed problem

7.3.1 Analysis of the continuous problem

Notation

Consider

$$\begin{cases} S(u)\,\sigma - \nabla u &= 0 \quad \text{in } \Omega, \\ -\operatorname{div}\,\sigma &= F \quad \text{in } \Omega, \\ u &= 0 \quad \text{on } \Gamma. \end{cases} \tag{7.4}$$

Define

$$X = [H(\operatorname{div},\Omega)]^2, \qquad Y = [L^2(\Omega)]^2, \qquad X_0 = \{\tau \in X : \ \operatorname{div}\tau = 0\},$$

with norms

$$\|\tau\|_X := \left(\|\tau\|_{0,\Omega}^2 + \|\operatorname{div}\tau\|_{0,\Omega}^2\right)^{1/2}, \qquad \|v\|_Y := \|v\|_{0,\Omega}.$$

On $[L^2(\Omega)]^{2\times 2}$ we use $(\sigma,\tau) = \int_\Omega \sigma : \tau$.

Green?s formula yields the mixed formulation: find $\sigma \in X$, $u \in Y$ such that

$$\begin{cases} \big(S(u)\,\sigma,\tau\big) + \big(u,\operatorname{div}\tau\big) &= 0 \qquad \forall \tau \in X, \\ \big(\operatorname{div}\sigma, v\big) + \big(F,v\big) &= 0 \qquad \forall v \in Y. \end{cases} \tag{7.5}$$

Let $a(\sigma,\tau) = (S(u)\sigma,\tau)$ and $b(\tau,u) = (\operatorname{div}\tau,u)$. Then a is continuous and coercive on X_0, and b is continuous (see Section 7.2.2).

7.3.1.1 Existence and uniqueness

Theorem 5.

$(\sigma, u) \in X \times Y$ solves (7.5) iff u solves

a. Existence:

$$\begin{cases} -\operatorname{div}\big(R(u)\,\nabla u\big) = F & \text{in } \Omega, \\ u = 0 & \text{on } \Gamma, \end{cases} \tag{7.6}$$

with $\sigma = R(u)\,\nabla u$.

Proof. As in the original text.

b. Uniqueness: We use the following lemma.

Lemma 3.

There exists $\beta_1 > 0$ such that

$$\sup_{\tau \in X,\, \tau \neq 0} \frac{(\operatorname{div}\tau, v)}{\|\tau\|_{0,\Omega}} \geq \beta_1 \|v\|_{0,4,\Omega} \qquad \forall v \in [L^4(\Omega)]^2.$$

Proof. By solving $-\Delta\omega = |v|^2 v$ with $\omega = 0$ on Γ and taking $\tau = -\nabla\omega$

Theorem 6.

If $\nabla u \in [L^4(\Omega)]^{2\times 2}$ and $\|F\|$ is sufficiently small, then (7.5) admits a unique solution (σ, u).

Proof. As in the original, using the Lipschitz continuity of S, the inf-sup estimate, and smallness of F.

We now turn to the discrete analysis. Section 7.3.2 treats the Picard linearization and its linear convergence; Section 7.3.3 treats discretization, well-posedness, and error bounds.

7.3.2 Linearization and convergence

We use Picard iteration: given u^0, for $k \geq 1$, find $(\sigma^k, u^k) \in X \times Y$ such that

$$\begin{cases} \big(S(u^{k-1})\,\sigma^k, \tau\big) + \big(\operatorname{div}\tau, u^k\big) &= 0 \qquad \forall \tau \in X, \\ \big(\operatorname{div}\sigma^k, v\big) + \big(F, v\big) &= 0 \qquad \forall v \in Y. \end{cases} \tag{7.7}$$

7.3.2.1 Well-posedness

Theorem 7.

For each $k \geq 1$, (7.7) admits a unique solution $(\sigma^k, u^k) \in X \times Y$.

Proof. Standard mixed theory; the bilinear form is coercive on the kernel and an inf-sup condition holds.

□

7.3.2.2 Linear convergence

Theorem 8.

Let (σ, u) solve (7.5). Under the hypotheses of Theorem 6, the sequence (σ^k, u^k) defined by (7.7) converges linearly to (σ, u).

Proof. As in the original derivation (inequalities (??)-(??)), yielding a contraction when $\tilde{M}\|F\|_{0,\Omega} < 1$.

□

7.3.3 Discrete problem and error estimate

7.3.3.1 Mixed discrete formulation

Let $\mathcal{T}_h$ be a regular triangulation of Ω, with $h = \max_T h_T$. Consider the lowest-order Raviart-Thomas element $RT_0(T) = [P_0(T)]^2 \oplus P_0(T)\,(x, y)$. Define

$$X_h = \{\tau_h \in X : \ \tau_{h|T} \in [RT_0(T)]^2 \ \forall T \in \mathcal{T}_h\}, \tag{7.8}$$

$$Y_h = \{v_h \in Y : \ v_{h|T} \in [P_0(T)]^2 \ \forall T \in \mathcal{T}_h\}, \tag{7.9}$$

$$Z_h = \{\tau_h \in X_h : \ \operatorname{div} \tau_h = 0\}. \tag{7.10}$$

For each $k \geq 1$, find $\sigma_h^k \in X_h$ and $u_h^k \in Y_h$ such that

$$\begin{cases} (S(u_h^{k-1})\,\sigma_h^k, \tau_h) + (\operatorname{div} \tau_h, u_h^k) & = 0 \qquad \forall \tau_h \in X_h, \\ (\operatorname{div} \sigma_h^k, v_h) + (F, v_h) & = 0 \qquad \forall v_h \in Y_h. \end{cases} \tag{7.11}$$

Using standard projection operators P_h and Π_h, one obtains the discrete inf-sup estimate

$$\sup_{\tau_h \in X_h,\, \tau_h \neq 0} \frac{(\operatorname{div} \tau_h, v_h)}{\|\tau_h\|_{0,\Omega}} \geq \beta_4 \|v_h\|_{0,4,\Omega} \qquad \forall v_h \in Y_h,$$

and thus well-posedness for (7.11).

7.3.3.2 A priori error estimate

Let (σ^k, u^k) and (σ_h^k, u_h^k) be the solutions of (7.7) and (7.11), respectively.

Theorem 9.

Assume $\sigma^k \in [H^1(\Omega)]^{2\times 2}$ and $u^k \in [W^{1,4}(\Omega)]^2$ for all $k \geq 1$, and take $u_h^0 = P_h u^0$. Then there exists $C^* > 0$ independent of h such that

$$\|\sigma^k - \sigma_h^k\|_{0,\Omega} + \|u^k - u_h^k\|_{0,4,\Omega} \leq C^*\, h.$$

Proof. Standard projection and stability arguments (see (??)-(??)).

□

Remark 7.3.1. *Since* $\|u^k - u_h^k\|_{0,\Omega} \leq |\Omega|^{1/4}\, \|u^k - u_h^k\|_{0,4,\Omega}$,

$$\|u^k - u_h^k\|_{0,\Omega} \leq C_1^*\, |\Omega|^{1/4}\, h.$$

7.4 Time-dependent problem

7.4.1 Continuous analysis

Notation

We consider

$$\begin{cases} M\partial_t u(x,t) - \operatorname{div}\sigma(x,t) & = F(x,t) \quad \text{in } \Omega_T := \Omega\times[0,T],\\ S(u(x,t))\,\sigma(x,t) - \nabla u(x,t) & = 0 \quad \text{in } \Omega_T,\\ u(x,t) & = 0 \quad \text{on } \Gamma\times[0,T],\\ u(x,0) & = u_0 \quad \text{in } \Omega. \end{cases} \tag{7.12}$$

We recall that

$$u = (h,\Phi_f)^T \in W(0,T)\times L^2(0,T,H_0^1(\Omega)),$$

with

$$W(0,T) := \{\, v\in L^2(0,T,H_0^1(\Omega)) \;:\; \partial_t v \in L^2(0,T,H^{-1}(\Omega))\,\}.$$

We seek $\sigma\in L^2(0,T,X)$ and $u\in E(0,T) := W(0,T)\times L^2(0,T,H_0^1(\Omega))$ such that, for all $v\in Y$, $\tau\in X$ and a.e. $t\in[0,T]$,

$$\begin{cases} \big(M\partial_t u, v\big) - \big(\operatorname{div}\sigma, v\big) & = \big(F,v\big),\\ \big(S(u)\,\sigma,\tau\big) + \big(u,\operatorname{div}\tau\big) & = 0,\\ u(\cdot,0) & = u_0 \qquad \text{in } \Omega. \end{cases} \tag{7.13}$$

Define $\|v\|_M^2 = \int_\Omega (Mv)\cdot v\,dx$, and the norms $\|\cdot\|_{L^2(0,T,Y_M)}$ and $\|\cdot\|_{L^\infty(0,T,Y_M)}$ as usual.

7.4.1.1 Existence and uniqueness

a. Existence:

Theorem 10.

$(\sigma, u) \in L^2(0,T,X) \times E(0,T)$ solves (7.13) iff u solves

$$\begin{cases} M\,\partial_t u - \operatorname{div}\big(R(u)\,\nabla u\big) = F, & \text{dans } \Omega_T, \\ u = 0, & \text{sur } \Gamma \times [0,T], \\ u(\cdot,0) = u_0, & \text{dans } \Omega. \end{cases}$$

with $\sigma = R(u)\,\nabla u$.

Proof. As in the stationary case, holding for a.e. t.

□

b. Uniqueness:

Theorem 11.

If $\nabla u \in [L^4(\Omega)]^{2\times 2}$ and $\|F\|$ is sufficiently small, then (7.13) admits a unique solution (σ, u).

Proof. As in the stationary case, employing the inf-sup estimate and smallness of F to conclude $u_1 = u_2$ and $\sigma_1 = \sigma_2$.

□

7.4.2 Linearized problem

We apply Picard iteration: given $u^0(x,t)$, for $k \geq 1$, find $(\sigma^k, u^k) \in L^2(0,T,X) \times E(0,T)$ such that

$$\begin{cases} \big(M\partial_t u^k, v\big) - \big(\operatorname{div}\sigma^k, v\big) & = \big(F, v\big) \quad \forall v \in Y, \\ \big(S(u^{k-1})\,\sigma^k, \tau\big) + \big(u^k, \operatorname{div}\tau\big) & = 0 \quad \forall \tau \in X, \\ u^k(\cdot,0) & = u_0 \quad \text{in } \Omega. \end{cases} \tag{7.14}$$

7.4.2.1 Well-posedness

Theorem 12.

For each $k \geq 1$, (7.14) admits a unique solution $(\sigma^k, u^k) \in L^2(0,T,X) \times E(0,T)$.

Proof. Uniqueness is standard; existence follows from equivalence with the linear parabolic problem

$$M\partial_t u^k - \operatorname{div}\big(R(u^{k-1})\nabla u^k\big) = F, \quad u^k|_{\Gamma\times[0,T]} = 0, \quad u^k(\cdot,0) = u_0,$$

and setting $\sigma^k = R(u^{k-1})\,\nabla u^k$; see, e.g., [65].

□

7.4.2.2 Linear convergence

Theorem 13.

Let (σ, u) be the unique solution of (7.13). Under the hypotheses of theorem 11, the sequence (σ^k, u^k) solving (7.14) converges linearly to (σ, u).

Proof. As in the original derivation, yielding a contraction for $\tilde{C}_4\|F\|_{L^\infty(0,T,Y)} < 1$.

□

7.4.3 Discrete problem and error estimates

We now analyze the discretization associated with (7.14), invoking the stationary results when needed.

7.4.3.1 Mixed discrete formulation

Define

$$W_h(0,T) := \{\, v_h \in L^2(0,T,V_h) \;:\; \partial_t v_h \in L^2(0,T,H^{-1}(\Omega)) \,\}, \quad E_h(0,T) := W_h(0,T)\times L^2(0,T,V_h$$

where $V_h = \{v_h \in L^2(\Omega) : v_{h|T} \in P_0(T),\ \forall T \in \mathcal{T}_h\}$. For each $k \geq 1$, find $(\sigma_h^k, u_h^k) \in L^2(0,T,X_h) \times E_h(0,T)$ such that

$$\begin{cases} \big(M\partial_t u_h^k, v_h\big) - \big(\operatorname{div}\sigma_h^k, v_h\big) = \big(F, v_h\big) & \forall v_h \in Y_h, \\ \big(S(u_h^{k-1})\,\sigma^k, \tau_h\big) + \big(u_h^k, \operatorname{div}\tau_h\big) = 0 & \forall \tau_h \in X_h, \\ u_h^k = u_{h,0} & \text{for } t = 0, \end{cases} \tag{7.15}$$

with $u_{h,0}$ an approximation of u_0 in Y_h.

Proposition 7.4.1. *Problem* (7.15) *admits a unique solution* $(\sigma_h^k, u_h^k) \in L^2(0,T,X_h) \times E_h(0,T)$.

Proof. Either via ODE reduction in time (projecting onto V_h) or via invertibility of appropriate projection operators; see [65, 27] and [119, Ch. 17].

□

Using the stationary solution operators T_h, S_h and the continuous operator T, one has T_h self-adjoint, positive semidefinite on $[L^2]^2$ and positive definite on Y_h, with

$$\|T_hF - TF\|_{0,\Omega} \le C^*h.$$

7.4.3.2 A priori error bounds

From theorem 9 we refine constants so that, for all $k \ge 1$,

$$\|u^k - u_h^k\|_{0,\Omega} \le C_1|\Omega|^{1/4}h\sum_{i=0}^{k}\left(|u^i|_{1,4,\Omega} + |\sigma^i|_{1,\Omega}\right),$$

$$\|\sigma^k - \sigma_h^k\|_{0,\Omega} \le C_2h\left(|\sigma^k|_{1,\Omega} + \sum_{i=0}^{k-1}(|u^i|_{1,4,\Omega} + |\sigma^i|_{1,\Omega})\right).$$

Theorem 14.

Let (σ^k, u^k) and (σ_h^k, u_h^k) be the solutions of (7.14) and (7.15), respectively, with $u_{h,0}^k = T_hLu_0^k$. Then, for a.e. $t \in [0,T]$,

$$\|u^k(t) - u_h^k(t)\|_{L^\infty(0,T,Y_M)} \le \hat{C}\,h\sum_{i=0}^{k}\left(\|u^i\|_{L^\infty(0,T,W^{1,4})} + \|\partial_t u^i\|_{L^2(0,T,W^{1,4})} + \|\sigma^i\|_{L^\infty(0,T,W^{1,2})} + \|\partial_t\sigma^i\|_{L^2(0,T,W^{1,2})}\right),$$

and

$$\|\sigma^k(t) - \sigma_h^k(t)\|_{L^\infty(0,T,W^{1,2})} \le \hat{C}\,h\Big(\sum_{i=0}^{k}\left(\|\sigma^i\|_{L^\infty(0,T,W^{1,2})} + \|\partial_t\sigma^i\|_{L^2(0,T,W^{1,2})}\right) + \sum_{i=0}^{k-1}\left(\|u^i\|_{L^\infty(0,T,W^{1,4})} + \|\partial_t u^i\|_{L^2(0,T,W^{1,4})}\right) + \|\partial_t u^k\|_{L^2(0,T,W^{1,4})}\Big),$$

with $\hat{C}$ independent of h.

Proof. Split the error using the elliptic projection $(\hat{\sigma}_h^k, \hat{u}_h^k)$, estimate projection and evolution parts separately, and apply Gronwall; see the original derivation ((4.1)-(4.5)).

Conclusion

This research aims to develop efficient mathematical approaches addressing two major application fields: the valuation of financial options and the modeling of saltwater intrusion in coastal aquifers. Although these topics belong to distinct domains, they share a common conceptual foundation based on modeling, the analysis of inverse problems, and the implementation of advanced numerical methods.

The first part, devoted to **financial mathematics**, focused on option pricing and volatility estimation. Beyond the historical perspective from Black–Scholes volatility was treated as a genuine inverse problem. Efficient algorithms and numerical solvers were developed, based on the Discontinuous Galerkin method in space coupled with a Runge–Kutta scheme (RKDG) in time. These tools enabled an accurate and stable reconstruction of the local volatility function without assuming its constancy. Numerical validations confirmed the speed, robustness, and competitiveness of the proposed approach compared with classical methods.

The second part, dedicated to **hydro-environmental modeling**, addressed the simulation of saltwater intrusion in coastal aquifers aa critical issue in regions facing increasing water stress and over exploitation of groundwater resources. By adopting a mixed sharp/diffuse-interface formulation, we established a well-posed mathematical framework ensuring mass conservation and a realistic representation of the transition between freshwater and saline zones. The Mixed Finite Element Method (MFEM) proved particularly suitable, providing accurate flux estimates, convergence guarantees, and robust numerical stability, even in complex physical configurations.

These two studies are linked by a common methodological thread centered on the **solution of inverse problems**. In finance, the inverse problem involves calibrating volatility parameters from market data; in hydrogeology, it consists in identifying source terms or hydraulic properties from partial measurements. In both cases, the challenges related to nonlinearity, instability, and ill-posedness are addressed through regularization techniques, stability analysis, and error-controlled numerical schemes.

From a scientific standpoint, this work highlights the capacity of shared mathematical principles—such as variational formulations, regularization, mixed finite elements, and discontinuous Galerkin methods—to transcend disciplinary boundaries. The parallel between financial and environmental modeling demonstrates the richness of this cross-fertilization: stochastic models developed in finance inspire approaches to uncertainty in hydrogeology, while conservative formulations derived from fluid mechanics enhance numerical robustness in financial models.

Bibliography

[1] A finite element method for hydraulic conductivity identification in a seawater intrusion problem. *Computers and Geosciences*, 33:860 – 874, 2007. (Cited on page 3.)

[2] Volatility articles. 2009-2012. `http://www.stock-options-made-easy.com/volatility-articles.html`. (Cited on page 18.)

[3] Mathematical analysis of a sharp–diffuse interfaces model for seawater intrusion . *Journal of Differential Equations*, 259:3803 – 3824, 2015. (Cited on page 3.)

[4] A nonlinear optimization method applied to the hydraulic conductivity identification in unconfined aquifers. *J Optim Theory Appl*, 183:705–730, 2019. (Cited on page 3.)

[5] E. Abarca, J. Ramírez, X. Sanchez-Vila, and C. Voss. Quasi-horizontal circulation cells in 3d seawater intrusion. *Journal of Hydrology*, 339:118–129, 06 2007. (Cited on page 3.)

[6] R. Aboulaich, H. Ben Ameur, and S.M. Lamarti. Volatility estimation via jump indicator. *Modern Applied Science*, 8(2):12–21, 2014. (Cited on pages 34, 52 and 54.)

[7] R. Aboulaich and I. Medarhri. Calibration of volatility: Coupling between alternative regularization strategy and dupire's equations. *Int. J. App. Math. Stat.*, 42(12):70–83, 2013. (Cited on page 28.)

[8] A. Abudawia, A. Mourad, J.H. Rodrigues, and C. Rosier. A finite element method for a seawater intrusion problem in unconfined aquifers. *Applied Numerical Mathematics*, 127:349 – 369, 2018. (Cited on page 3.)

[9] Y. Achdou, O. Bokanowski, and T. Lelièvre. Partial differential equation in finance. *Rapport de Recherche - CERMICS*, (363), 2007. (Cited on page 25.)

[10] Y. Achdou and O. Pironneau. *Computational methods for option pricing.* SIAM. Philadelphia, PA, 19104–2688, 2005. (Cited on pages 12, 15, 16 and 17.)

[11] Y. Achdou and N. Tchou. Variational anaysis for the Black and Scholes equation with stochastic volatility. *Mathematical Modelling and Numerical Analysis*, 3:373 – 395, 2002. (Cited on page 89.)

[12] P. Ackerer and A. Younes. Efficient approximations for the simulation of density driven flow in porous media. *Advances in Water Resources*, 31:15–27, 2008. (Cited on page 3.)

[13] A. Ait Hammou Oulhaj. Numerical analysis of a finite volume scheme for a seawater intrusion model with cross-diffusion in an unconfined aquifer. *Numerical Methods for Partial Differential Equations*, 34, 2017. (Cited on page 3.)

[14] M. Avellaneda, C. Friedman, R. Holmes, and D . Samperi. Calibrating volatity surfaces via relative entropy minimization. *Applied hhh Mathematical*, 4(4), 1997. (Cited on pages 27 and 28.)

[15] L. Bachelier. Théorie de la spéculation. *thèse, Ann. Sci. de l'École Norm. Sup*, 3(17):21–86., 1900. (Cited on page 2.)

[16] M. BATCHI, J. Al KARKOURI, M. El MAAQILI, and I. FENIJIRO. Cartographie à la vulnérabilité à l'intrusion marine de l'aquifère côtier de Mnasra (Littoral du Gharb, Maroc-Nord-Occidental. *Marine Sciences and Coastal Research*, 1, 2014. (Cited on page 2.)

[17] J Bear. *Dynamics of Fluids in Porous Media.* American Elsevier, 1972. (Cited on pages 2, 29 and 31.)

[18] J. Bear and A.H.D. Cheng. *Modeling Groundwater Flow and Contaminant Transport.* Springer, 2010. (Cited on page 2.)

[19] J. Bear, A.H.D. Cheng, S. Sorek, D. Ouazar, and I. Herrera. *Seawater intrusion in coastal aquifer : Concept, methods and pratices.* Klewer Academic Pub, 1999. (Cited on pages 2 and 31.)

[20] J. Bear and A. Verruijt. *Modelling groundwater flow and pollution.* D. Reidel Publishing Company, 1987. (Cited on page 2.)

[21] H. Benameur, G. Chavent, and J. Jaffré. Refinement and coarsening indicators for adaptive parametrization: application to the estimation of hydraulic transmissivities. *Inverse Problems*, 18(3):229 – 263, 2002. (Cited on page 52.)

[22] F. Black and M. Scholes. The pricing of options and corporate liabilities. *Journal of Political Economy*, 81(3):637–659, 1973. (Cited on pages 2 and 27.)

[23] D. Boffi, F. Brezzi, and M. Fortin. *Mixed Finite Element Methods and Applications.* Springer-Verlag, (2013). (Cited on page 3.)

[24] O. Bokanowski, Y. Cheng, and Shu C.W. A Discontinuous Galerkin Scheme for Front Propagation with Obstacles. *Numerische Mathematik*, 126(1):1–31, 2014. (Cited on page 86.)

[25] J.F. Bonnans, J.M. Cognet, and S. Volle. Estimation de la volatilité locale d'actifs financiers par une méthode d'inversion numérique. *Rapport de recherche INRIA*, (4648), 2002. (Cited on pages 2, 3, 28, 34 and 51.)

[26] I. Bouchouev and V. Isakov. Uniqueness, stability and numerical methods for the inverse problem that arises in financial markets. *Inverse Problems*, 15:R95–R116, 1999. (Cited on pages 27 and 28.)

[27] L. Boulaajine, M. Farhloul, and L. Paquet. A priori error estimation for the dual mixed finite element method of the elastodynamic problem in a polygonal domain, i. *Journal of Computational and Applied Mathematics*, 231:447–472, 2009. (Cited on page 116.)

[28] P. Brandimarte. Numerical methods in finance and economics. *John Wiley & Sons, Inc, Second edition*, 2006. (Cited on page 15.)

[29] F. Brezzi and M. Fortin. *Mixed and Hybrid Finite Element Methods*. Springer, (1991). (Cited on page 3.)

[30] A. Charpentier. Méthodes numériques en finance. 2006/2007. `http://ww.crest.fr/pageperso/charpent/charpentFR.htm`. (Cited on pages 16 and 17.)

[31] G. Chavent and J. E. Roberts. A unified physical presentation of mixed, mixed-hybrid finite elements and standard finite difference approximations for the determination of velocities in waterflow problems. *Advances in Water Resources*, 14:329–348, 1991. (Cited on page 3.)

[32] Y. Cheng and C.-W. Shu. A discontinuous Galerkin finite element method for directly solving the Hamilton-Jacobi equations. *Journal of Computational Physics*, 223:398–415, 2007. (Cited on pages 60, 68 and 69.)

[33] Laurence Cherfils, Catherine Choquet, and Moussa Diedhiou. Numerical validation of an upscaled sharp–diffuse interface model for stratified miscible flows. *Mathematics and Computers in Simulation*, 137:246–265, 2017. (Cited on page 3.)

[34] C. Chiarella, M. Craddock, and N. El-Hassan. The calibration of stock option pricing models using inverse problem methodology. *School of finance and Econimics,University of Technology Sydney*, 2007. (Cited on pages 2, 3, 28, 34, 39, 41, 42, 43, 46, 47 and 57.)

[35] C. Choquet. Parabolic and degenerate parabolic models for pressure-driven transport problems. *Mathematical Models & Methods in Applied Sciences - M3AS*, 20, 04 2010. (Cited on page 3.)

[36] C. Choquet, M. Diedhiou, and C. Rosier. Derivation of a sharp-diffuse interfaces model for seawater intrusion in a free aquifer. numerical simulations. *Mathematics and Computers in Simulation*, 76:138–158, 2016. (Cited on pages 3, 4, 28 and 30.)

[37] C. Choquet, J. Li, and C. Rosier. Global existence for seawater intrusion models: Comparison between sharp interface and sharp-diffuse interface approaches. *Electronic Journal of Differential Equations*, pages 1–27, 2015. (Cited on page 3.)

[38] C. Choquet, J. Li, and C. Rosier. Numerical analysis for a seawater intrusion problem in a confined aquifer. *Mathematics and Computers in Simulation*, 118:2–16, 2015. (Cited on page 3.)

[39] C. Choquet, J. Li, and C. Rosier. Uniqueness for cross-diffusion systems issuing from seawater intrusion problems. *Electronic Journal of Differential Equations*, pages 1–22, 2017. (Cited on pages 3 and 94.)

[40] P. Ciarlet. *The Finite Element Methods for Elliptic Problem.* (1978). (Cited on page 3.)

[41] P. G. Ciarlet. *Lectures on the Finite Element Method*, volume 49. Tata Institute of Fundamental Research, Lectures on Mathematics and Physics, Bombay, 1975. (Cited on pages 16 and 72.)

[42] B. Cockburn. (Cited on pages 60 and 71.)

[43] B. Cockburn. Discontinuous Galerkin Methods. *School of Mathematics, University of Minnesota. Available at.* `http://commath.files.wordpress.com/2008/10/discontinuous_galerkin_methods_cockburn.pdf`. (Cited on page 60.)

[44] B. Cockburn, S. Hou, and C. W. Shu. The Runge-Kutta local projection discontinuous Galerkin finite element method for conservation laws IV: the multidimensional case. *Mathematics of computation*, 54:545–581, 1990. (Cited on page 60.)

[45] B. Cockburn, S. Y. Lin, and C. W. Shu. TVB Runge-Kutta local projection discontinuous Galerkin finite element method for conservation laws III: One dimentional systems. *Journal of computation Physics*, 84:90–113, 1989. (Cited on pages 17, 60 and 69.)

[46] B. Cockburn and C. W. Shu. TVB Runge-Kutta local projection Discontinuous Galerkin finite element method for conservation laws II: general framework. *Mathematics of computation Physics*, 52:411–435, 1989. (Cited on page 60.)

[47] B. Cockburn and C. W. Shu. The Runge-Kutta local projection P^1 discontinuous-Galerkin finite element method for scalar conservation laws. *Mathematical Modelling and Numerical Analysis (M^2AN)*, 25:337–361, 1991. (Cited on page 60.)

[48] B. Cockburn and C.-W. Shu. The Local Discontinuous Galerkin Method for time-dependent convection- diffusion systems. *SIAM. Numer. Anal*, 35:2240–2463, 1998. (Cited on pages 17 and 63.)

[49] B. Cockburn and C. W. Shu. The Runge-Kutta discontinuous-Galerkin method for conservation laws V: multidimensional systems. *Journal of Computation Physics*, 141, 1998. (Cited on page 60.)

[50] T. Coleman, Y. Li, and A. Verma. Reconstructing the unknown volatility function. *Journal of computational Finance*, 2(3):77–102, 1999. (Cited on pages 2 and 27.)

[51] R. Cont and P. Tankov. *Financial modeling with jump processes*, volume 1 edition of *Chapman & Hall/CRC Financial Mathematics*. Boca Raton, FL, 2004. (Cited on page 25.)

[52] R. Courant. Variational methods for the solution of problems of equilibrium and vibrations. *Bultin of the Americain Mathematical Society*, 49:1–23, 1943. (Cited on page 16.)

[53] J. C. Cox, S. Ross, and M. Rubinstein. Option pricing : a simplified approach. *Journal of Finance Economics*, 7:229 – 263, 1979. (Cited on page 13.)

[54] S. Crépey. Calibration of the local volatility in a trinomial tree using tikhonov regularization. *Stavks.iop.org*, (IP/19/91), 2002. (Cited on pages 27 and 28.)

[55] S. Crépey. Tikhonov regularization and calibration of a local volatility in finance-stability, convergence and convergence rates issues,. *CMAP Internal Research Report*, (474), 2002. (Cited on page 27.)

[56] A. De Cezaro and J. P. Zubelli. On a parabolic inverse problem arising in quantitative finance: Convex and iterative regularization. *Instituto Nacional de Matematica Pure e aplicadia*, 2010. (Cited on page 5.)

[57] E. Derman and I. Kani. Riding a smile. *Risk,vol"*, 17(2):32–39, 1994. (Cited on pages 23 and 27.)

[58] Ern A. Di Pietro, D. A. *Mathematical Aspects of Discontinuous Galerkin Methods*, volume 69. SMAI Mathématiques et Applications, Springer, Verlag Berlin Heidelberg, 2012. (Cited on page 17.)

[59] J. Duan and J. Wei. Pricing foreign-currency and options under GARCH. *Journal of Derivatives*, 7(1):33–51, 1999. (Cited on page 26.)

[60] B. Dupire. Pricing with smile. *Risk, vol*, 7(1):18–20, 1994. (Cited on pages 23, 27 and 28.)

[61] L. J. Durlofsky. Accuracy of mixed and control volume finite element approximations to Darcy velocity and related quantities. *Water Resources Research*, 1994. (Cited on page 3.)

[62] H. Egger and H.W. Engl. Tikhonov regularization applied to the inverse problem of option pricing: convergence analysis and rates. *Inverse Problems*, 21(3):1027–1045, 2005. (Cited on page 28.)

[63] M. El Alaoui Talibi and M.H. Tber. Existence of solutions for a degenerate seawater intrusion problem. *Electronic Journal of Differential Equations*, 2005. (Cited on page 3.)

[64] M. El Mokhtar, M. Chibout, M. Kili, and al. et. Évaluation de l'intrusion saline dans la nappe de Foum El Oued, province de Laâyoune, Maroc. *Bulletin de l'Institut Scientifique, Rabat, Section Sciences de la Terre*, 40:53–69, 2018. (Cited on page 2.)

[65] M. Farhloul, R. Korikache, and L. Paquet. The dual mixed finite element method for the heat diffusion equation in a polygonal domain, i. *Functional Analysis and Evolution Equations*, pages 239–256, 2008. (Cited on pages 115 and 116.)

[66] A. Fatullayev. Numerical solution of the inverse problem of determining an unknown source term in a heat equation. *Mathematics and Computers in Simulation*, 58:247 – 253, 2002. (Cited on page 3.)

[67] A. Fatullayev. Numerical solution of the inverse problem of determining an unknown source term in a two-dimensional heat equation. *Applied Mathematics and Computation*, 152:659–666, 2004. (Cited on page 3.)

[68] J.P. Fouque, G. Papanicolaou, and K. R. Sircar. *Derivatives in financial markets with stochastic volatility*, volume 1 edition. Cambridge University Press, Cambridge, 2000. (Cited on page 24.)

[69] V. Girault and P.A. Raviart. *Finite Element Methods for Navier-Stokes Equations : Theory and Algorithms*. Springer-Verlag, (1986). (Cited on page 3.)

[70] J. Glover and A. Montaz. Using radial basis functions to construct local volatility surfaces. *Applied Mathematics and Computation*, 217(9), 2011. (Cited on page 28.)

[71] E. Gobet, G. Pagès, and M. Yor. Mathématiques et finance. *Proba.jussieu.fr*. (Cited on page 1.)

[72] T. Graf and R. Therrien. Variable-density groundwater flow and solute transport in irregular 2d fracture networks. *Advances in Water Resources*, 30:455–468, 2007. (Cited on page 3.)

[73] J. Hadamard. Lectures on cauchy's problem in linear partial differential equations. *Dover Publications*, 1953. (Cited on pages 26 and 27.)

[74] T.K. Rodriguez G. Hansen, C. Jensen. An adaptive pruning algorithm for the discrete l-curve criterion. 198:483–492. (Cited on page 101.)

[75] Warburton T. Hesthaven, J. S. *Nodal Discontinuous Galerkin Methods. Algorithms, Analysis, and applications*, volume 54. Springer Science + Business Media, LLC, New York, 2008. (Cited on page 17.)

[76] S. Heston. A closed-form solution for options with stochastic volatility with applications to bond and currency options. *Review of Financial Studies.*, 6(2):327–344, 1993. (Cited on page 24.)

[77] R. Hofste, S. Kuzma, S. Walker, E.H. Sutanudjaja, and et. al. "aqueduct 3.0: Updated decisionrelevant global water risk indicators. *Technical Note. Washington, DC: World Resources Institute. Available online at: https://www.wri.org/publication/aqueduct-30.* (Cited on page 2.)

[78] J. Hull and A. White. The pricing of options on assets with stochastic volatility. *Journal of Finance.*, 42:281–300, 1987. (Cited on page 24.)

[79] J.C. Hull. *Options Futures And Other Derivatives*, volume 7 of *Pearson Education International*. Prentice Hall, Toronto, 2008. (Cited on pages 1, 6, 7, 8, 9, 12, 15, 16, 17, 22 and 50.)

[80] K. Itô. Stochastic integral. *Proc. Imp. Acad. Tokyo*, 20:519–524, 1944. (Cited on page 2.)

[81] N. Jakson, E. Süli, and S. Howison. Computation of deterministic volatility surfaces. *Journal Comp. Finance*, 2:5–32, 1998. (Cited on pages 2 and 28.)

[82] M. Jazar and R. Monneau. Formal derivation of seawater intrusion models. 03 2011. (Cited on page 3.)

[83] Y. Jerbi. Evaluation des options et gestion des risques financier par les réseaux de neurones et par les modèles à volatilité stochastiques. *Thèse de doctorat*, 2006. (Cited on pages 1 and 8.)

[84] G. S. Jiang and C. W. Shu. On cell entropy inequality for discontinuous Galerkin method for a scalar hyperbolic equation. *Mathematics of computation*, 62:531–538, 1994. (Cited on pages 68, 69 and 70.)

[85] E. F. Kaasschieter. Mixed finite elements for accurate particle tracking in saturated groundwater flow. *Advances in Water Resources*, 18:277–294, 1995. (Cited on page 3.)

[86] N. El Karaoui and E. Gobet. Les outils stochastiques des marché financière. *Les Editions de l'École Polytechnique*, 2011. (Cited on pages 1 and 8.)

[87] M. Kern. Problèmes inverses : aspects numériques. *Engineering school. 1999 à 2002, École supérieure d'ingénieurs Léonard de Vinci*, 162:138.pp, 2002. (Cited on page 3.)

[88] A. Kirsh. *An Introduction to the Mathematical Theory of Inverse Problems*, volume 120 of *Applied Mathematical Sciences*. Springer, Germany, 2011. (Cited on pages 38, 39, 40 and 41.)

[89] O. Kolditz, R. Ratke, H.G. Diersch, and W. Zielke. Coupled groundwater flow and transport: 1. verification of variable density flow and transport models. *Advances in Water Ressources*, 21, 1998. (Cited on page 3.)

[90] A. N. Kolmogorov. *English translation: Foundations of Probability Theory*. Chelsea Publ. Co, 1950. (Cited on pages 2 and 34.)

[91] M. Kulbay, B. Mukanova, and C. Sebu. Identification of separable sources for advection-diffusion equations with variable diffusion coefficient from boundary measured data. *Inverse Problems in Science and Engineering*, 25:2:279–308, 2017. (Cited on pages 3, 91, 92, 93, 95, 97 and 103.)

[92] R. S. Kunwar, D. Valsan, and A. Sahasrabudhe. Inverstment banking products. *L&T Infotech.* (Cited on page 8.)

[93] R. Lagnado and S. Osher. A technique for calibrating derivative security pricing models: Numerical solution of an inverse problem. *The journal of computational Finance*, 1:13–25, 1997. (Cited on pages 2, 3, 27, 28, 34, 42, 47 and 57.)

[94] P. Lesaint and P. A. Raviart. One a finite element method for solving the neutron transport equation. *In Mathematical Aspects of Finite Elements in Partial Differential Equations (ed. C. de Boor). Academic Press, New York*, 1974. (Cited on page 60.)

[95] Randal J. LeVeque. Finite Volume Methods for Hyperbolic Problems. *Cambridge University Press*, 2002. (Cited on page 17.)

[96] X. Lian and H. Rui. A discontinuous Galerkin method combined with mixed finite element for seawater intrusion problem. *Journal of Systems Science & Complexity*, 23:830–845, 2010. (Cited on page 3.)

[97] B. Lu and A. Muzzucato. Monte Carlo simulations and option pricing. *Undergraduate Mathematics Departement, University Parck*, 2011. (Cited on page 12.)

[98] T. Luo, R. Young, and P. Reig. Aqueduct projected water stress country rankings. *Technical Note. Washington, D.C.: World Resources Institute. Available online at: www.wri.org/publication/aqueduct-projected-water-stresscountry-rankings.* (Cited on page 2.)

[99] P. Marion, K. Najib, and C. Rosier. Numerical simulations for a seawater intrusion problem in a free aquifer. *Applied Numerical Mathematics*, 75:48–60, 2014. (Cited on page 3.)

[100] R. Merton. Theory of rational option pricing. *Bell Journal of Economics and Management Science*, 4:141 – 183, 1973. (Cited on page 2.)

[101] K. W. Morton and D. F. Mayers. Numerical Solution of Partial Differential Equations: An Introduction. *Cambridge University Press, Second edition*, 2005. (Cited on page 15.)

[102] R. Mose, S. Pascal, P. Ackerer, and G. Chavent. Application of the mixed hybrid finite element approximation in a groundwater flow model: Luxury or necessity? *Water Resources Research*, 30:3001–3012, 1994. (Cited on page 3.)

[103] B. Mukanova and V. Romanov. Inverse source problem for wave equation and gpr data interpretation problem. *Eurasian Journal of Mathematical and Computer Applications*, 4, 2016. (Cited on page 3.)

[104] K. Najib and C. Rosier. On the global existence for a degenerate elliptic–parabolic seawater intrusion problem. *Mathematics and Computers in Simulation*, 81:2282–2295, 2011. (Cited on page 3.)

[105] B. Øksendal and A. Sulem. *Applied stochastic control of jump diffusions.*, volume 2nd Edition of *Universitext*. Springer-Verlag, New York, 2007. (Cited on page 88.)

[106] M. Parkinson. The externe value method for estimating the variance of the rate of return. *Journal of business*, 53, 1980. (Cited on page 19.)

[107] Vincent E. A. Post. Annotated translation of bnota in verband met de voorgenomen putboring nabij amsterdam (note concerning the intended well drilling near amsterdam) by j. drabbe and w. badon ghijben (1889). *Hydrogeology Journal*, 26:1771–1788, 2018. (Cited on page 2.)

[108] P. Raviart and J. Thomas. A mixed finite element method for second order elliptic problems. *Mathematical Aspects of Finite Element Methods, Lecture Notes in Mathematics*, 606:292–315, 1977. (Cited on page 3.)

[109] W.H. Reed and T. R. Hill. Triangular mesh methods for the neutron transport equation. *Tech. Report LA-UR-73-479, Los Alamos Sciendtific Laboratory*, 1973. (Cited on page 60.)

[110] T. E. Reilly and A. S. Goodman. Quantitative analysis of saltwater-freshwater relationships in groundwater systems—a historical perspective. *Journal of Hydrology*, 80:125–160, 1985. (Cited on page 2.)

[111] T. E. Reilly and A. S. Goodman. Analysis of saltwater upconing beneath a pumping well. *Journal of Hydrology*, 89:169–204, 1987. (Cited on page 2.)

[112] M. Richardson. Numerical methods for option pricing. 2009. `https://people.maths.ox.ac.uk/richardsonm/OptionPricing.pdf`. (Cited on pages 15, 16 and 17.)

[113] M. Rubinstein. Implied binomial trees. *Journal of Finance*, (69):771–818, 1994. (Cited on page 27.)

[114] A. Sasane. *Optimisation in function spaces.* London School of Economics. London. (Cited on page 43.)

[115] L. O. Scott. Pricing stocks options in a jump-diffusion model with stochastic volatility and rates applications of Fourier inversion methods. *Mathematical Finance (An International Journal of Mathematics, Statistics and Financial Econmics.)*, 7(4):413–426, 1997. (Cited on page 25.)

[116] C.-W. Shu. Discontinuous Galerkin Methods : General approach and stability. *Advenced courses in Mathematics CRM Barcelona. Available at*, 2009. `http://www.dam.brown.edu/scicomp/media/report_files/BrownSC-2008-08.pdf`. (Cited on pages 60 and 68.)

[117] E. M. Stein and J. C. Stein. Stock price distributions with stochastic volatility : An analytic approach. *Review of Financial Studies*, 4:727 – 752., 1991. (Cited on page 24.)

[118] M. H. Tber, E. M. El Alaoui, and D. Ouazar. Parameters identification in a seawater intrusion model using adjoint sensitive method. *Mathematics and Computers in Simulation*, 77:301 – 312, 2008. (Cited on page 3.)

[119] V. Thomée. *Galerkin Finite Element Methods for Parabolic Problems* . Springer Series in Computational Mathematics 25, Springer-Verlag, 1997. (Cited on page 116.)

[120] A.N. Tikhonov. Regularization of incorrectly posed problems. *Soviet Mathematics*, 4:1624–1627, 1963. (Cited on pages 27 and 42.)

[121] E. F. Toro. Riemann solvers and numerical methods for fluid dynamics. *Spinger-Verlag*, 1999. (Cited on page 17.)

[122] A. Verruijt. A note on the ghyben-herzberg formula. *Hydrological Sciences Journal-journal Des Sciences Hydrologiques - HYDROLOG SCI J*, 13:43–46, 1968. (Cited on page 2.)

[123] A.T. Vetterling W.T. Flannery B. P. William, H.P. Saul. *Numerical recipes in Fortran 77: the art of scientific computing.* Volume 1 of Fortran Numerical Recipes, Cambridge University Press, 1992. (Cited on page 100.)

[124] H. A. Windcliff, P.A. Forsyth, and K. R. Vetzal. Numerical methods and volatility model for valuing cliquet options. *Applied Mathematical Finance*, 13(4):353–386, 2006. (Cited on page 28.)

[125] A. Younes, R. Mose, P. Ackerer, and G. Chavent. Une résolution par les éléments finis mixtes à une inconnue par maille. *Comptes Rendus de l'Académie des Sciences - Series I - Mathematics*, 328:623–626, 1999. (Cited on page 3.)

[126] A. Younsi, E. Lhadi, and J. Manya. Mise en évidence de l'intrusion marine de la nappe d'El Jadida, Maroc. *Hydrochemistry (Proceedings of the Rabat Symposium)*, 244, 1997. (Cited on page 2.)

[127] J.S. Zhang, J. Zhu, and D.P. Yang. A Combined Discontinuous Galerkin Method for Saltwater Intrusion Problem with Splitting Mixed Procedure. *Advances in Applied Mathematics and Mechanics*, 9:651–666, 2017. (Cited on page 3.)

[128] J.S. Zhang, J. Zhu, D.P. Yang, and H. Guo. A characteristic splitting mixed finite element method for three-dimensional saltwater intrusion problem. *Journal of Nonlinear Science and Applications*, 9:5806–5821, 2016. (Cited on page 3.)

[129] M. Zhang and Shu C.W. An analysis of threre different formulations of the Discontinuous Galerkin Method for diffusion Equations. *Mathematical Models and Methods in Applied Sciences*, 13:395–413, 2003. (Cited on page 60.)

[130] O.C. Zienkiewicz and R.L. Taylor. *An Introduction to the Mathematical Theory of Inverse Problems*, volume 1 of *Applied Mathematical Sciences*. Butterworth-Heinemann, Oxford, fifth edition, 2000. (Cited on page 16.)

[131] L. Zouhri, E. Toto, E. Carlier, and T.H. Debieche. Salinité des ressources en eau: intrusion marine et interaction eaux–roches (maroc occidental). *Hydrological Sciences Journal-journal Des Sciences Hydrologiques - HYDROLOG SCI J*, 55:1337–1347, 2010. (Cited on page 3.)

APPENDIX A

Appendix

A.1 The TVD-Runge-Kutta time discretization

To discretize our ODE system in time, we use the TVD Runge Kutta time discretization introduced by Cockburn & Shu (1989). The equation 4.63 can be rewritten in the form below:

$$\begin{cases} \dfrac{u^{n+1} - u^n}{\tau} = L(r^n), \\ r^n = L_1(u^n) \end{cases} \tag{A.1}$$

Where,

$$L(r^{n,i}) = -(Ar^{n,i} + Br^{n,i-1}), \text{ and } L_1(u^{n,i}) = -(Du^{n,i} + Cu^{n,i+1})$$

- Algorithm for Runge-Kutta Method :

  Thus, if $(t^n)_{n=0}^{n=N}$ is a partition of [0,T] and $\tau^n = t^{n+1} - t^n; n = 0, ..., N-1$;

  Our time-marching algorithm reads as follows:

  - Set $u^0(x) = u_0(x)$;
  - For n =0, ...,N-1, compute u^{n+1} from u^n, and r^n, as follows :

  1. Set $u^0 = u^n$;
  2. for m=1,...,k+1 compute the intermediate functions:

  $$r^m = L_1(u^m)$$

  $$u^m = \{\sum_{k=0}^{m-1} \alpha_{ml} u^l + \beta_{ml} \tau L(r^{(l)})\};$$

  where α and β is the parameters of Runge-kutta Method

  3. set u^{n+1}=u^n.

- standard RK method We will use the following standard RK methods for a system of the form:

1. Euler scheme: r^n is calculate in equation A.1

$$u^{n+1} = u^n + \tau L(r^n) \ \equiv: \ S(t_n, r^n).$$

2. TVD-RK2 scheme:

$$u^{n+1} = \frac{1}{2}(u^n + S(t_{n+1}, S(t_n, r^n))).$$

3. TVD-RK3 scheme: with $t_{n+\frac{1}{2}} = t_n + \frac{\tau}{2}$:

$$u^{n,1} := S(t_{n+1}, S(t_n, u^n)); \tag{A.2a}$$

$$u^{n,2} := \frac{3}{4}u^n + \frac{1}{4}u^{n,1}, \tag{A.2b}$$

$$u^{n+1} := \frac{1}{3}u^n + \frac{2}{3}S(t_{n+\frac{1}{2}}, u^{n,2}). \tag{A.2c}$$

FSC
www.fsc.org
MIX
Papier | Fördert gute Waldnutzung
FSC® C083411

Kolibri 360 GmbH
Ferdinand-Jühlke-Straße 7
99095 Erfurt, Deutschland
produktsicherheit@kolibri360.de

Druck:
CPI Druckdienstleistungen GmbH
im Auftrag der
Kolibri 360 GmbH
Ferdinand-Jühlke-Str. 7
99095 Erfurt

Nine Yana